PREFACE

CONTENTS

The Earth is a tiny dot swamped by the
vast expanse of space. It is one of nine
planets that orbit the Sun, making up
our Solar System. Th
giving heat and light
100 billion (100,000
the Milky Way galaxy
Way is just one of bil
the universe.

This book is a brie
time, starting at the u
dense speck from wh
originated aeons befo
formed. Hurtling thro
formation of the Sola
explore our planet's p
detail, including the c
land, sea, atmosphere
until we reach today's
Earth supporting at le
living species.

Tackling a journey
itself, we can do no n
fleetingly every few n
consider some of the
Earth's turbulent hist

To aid the reader a
indicating the span of
those pages, appears

and reefs, with a general interest in
biogeography and ecology.

This book has been written to
complement the *From the Beginning*
exhibition in the Earth Galleries at
the Natural History Museum, London.

Front cover: View of Earth with Sunburst.

THE BIG BANG – BIRTH OF THE UNIVERSE

Fig 1 Edwin Hubble 1889-1953.

The origin of the Universe is among the most debated philosophical and scientific questions of our time. Our understanding of the birth of time began back in 1929 when the astronomer Edwin Hubble (fig 1) discovered that the Universe is currently expanding. All the stars and galaxies are gradually moving away from each other as space stretches evenly between them – think of the stars and galaxies as points on a balloon that is steadily being blown up, so that the distance between the points is gradually increasing. By using this concept, running time backwards and shrinking the Universe to its starting point, scientists have been able to work out when the Universe began and to imagine what it must have been like when it formed.

Although the exact figure is still being debated, many scientists agree that everything, including all matter and time itself, began about 13,700 million years (Ma) ago, when a tiny fireball of infinite density and heat exploded. During this event, known as the Big Bang (fig 2), scorching heat flew out in all directions, forming a hot soup of sub-atomic particles. Meanwhile the fireball was expanding, perhaps even faster than the speed of light.

At one minute protons and neutrons fused to form helium nuclei (each nucleus is composed of two protons and two neutrons). After about 300,000 years the temperature had dropped enough to allow electrons to orbit the

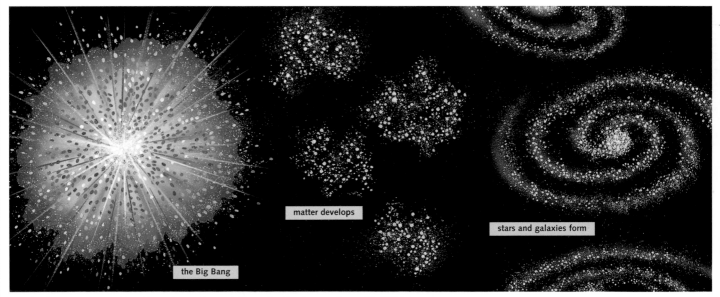

the Big Bang

matter develops

stars and galaxies form

Fig 2 The Universe is thought to have formed from the Big Bang about 13,700 Ma ago.

nuclei, making atoms. An expanding cloud of hydrogen and helium gas then formed.

The Universe probably remained dark until several billion years later when matter began to contract and clump together, creating galaxies and stars (fig 3). A star is a vast ball of hydrogen that is gradually converted to helium – a reaction called nuclear fusion – releasing enormous quantities of heat and light energy during the process. Our galaxy, the Milky Way, formed about 10,000 Ma ago. At its centre is a flattened disc with a central bulge of ancient stars from which four spiralling arms radiate. The whole structure is enveloped by a halo of dark matter and clusters of old stars.

Although the scenario for the Big Bang was put forward in the 1940s, it was not until the 1960s that direct evidence for an ancient explosion of heat was discovered. A uniform blanket of heat surrounding the Earth, called microwave background radiation, was detected, and this led scientists to conclude that it must be of cosmic origin and left over from the early rapid expansion of heat and light.

There is compelling evidence to suggest that the Big Bang model explains the origin of the Universe. But there are still unanswered questions, such as what caused the Big Bang in the first place, and what, if anything, existed before it took place?

Fig 3 The Eagle Nebula – columns of dust and gas from which stars are born.

THE SOLAR SYSTEM FORMS

Around 4560 Ma ago the history of planet Earth was just beginning. Out of a swirling cloud of gas and dust – called a nebula – our Solar System was born (fig 4). Inside the nebula, some of the matter had previously been drawn by gravity into a cloud-surrounded protostar which heated up in the process. The cloud flattened into a spinning disc and, as ever more material was concentrated at the centre, the protostar spontaneously ignited to form our star, the Sun. The Sun's core temperature rose to several million degrees centigrade (°C), triggering nuclear fusion and releasing the enormous amounts of energy that make the Sun a giant fireball.

The leftover gas and dust particles continued to orbit the new Sun, clumping together to form planetesimals, which were mini-planets of varying sizes reaching up to several thousand kilometres in diameter. Repeated collisions of planetesimals gradually formed complete planets, eventually creating the Solar System as we know it today.

Encircling the outer limits of the Solar System is the Oort cloud, containing the icy, rocky lumps known as comets, which orbit the Sun at random angles. Next comes the Kuiper Belt, a broad region of small objects made of rock and ice. Pluto, the outermost planet, may be a Kuiper Belt object. Then there are the two ice giants, Neptune and Uranus, whose distance from the Sun keeps them chilled.

Just close enough to the Sun to feel its warmth are the two gas giants, Saturn and Jupiter (fig 5). Each has a small rock-ice core swathed by a large envelope of hydrogen and helium, which were captured once the newly forming planets grew large enough to exert a strong gravitational pull. In between the orbits of Jupiter and Mars lies a band of orbiting rocky lumps called the asteroid belt. Asteroids, or fragments of them, are occasionally

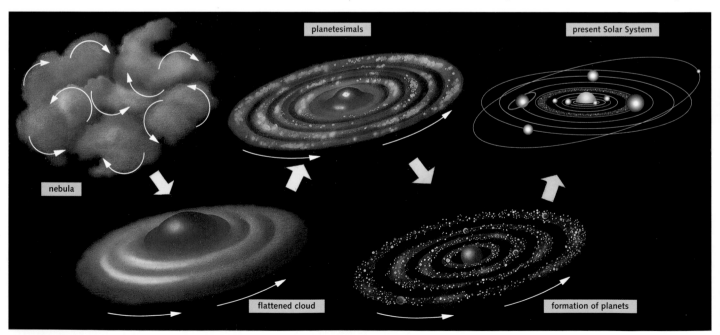

nebula

planetesimals

present Solar System

flattened cloud

formation of planets

Fig 4 Formation of our Solar System 4560 Ma ago.

knocked out of orbit, and the ones that cross Earth's path and survive the journey through our atmosphere to land on the surface are called meteorites.

Closer still to the Sun are the four rocky planets, Mars, Earth, Venus and Mercury (figs 6–8), each of which has an iron core and an outer layer composed of mainly silicate rocks. In contrast to the gas giants, the rocky planets may not have grown large enough to capture the early gases from the solar nebula before they dispersed, or these planets may have formed later in a gas-free environment.

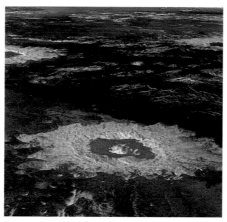

Fig 6 Lava flows and volcanic craters cover the surface of Venus where temperatures soar due to a runaway greenhouse effect.

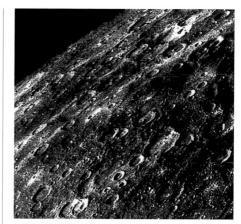

Fig 7 The cratered surface of Mercury is the result of continued pounding by meteorites soon after the planet's formation.

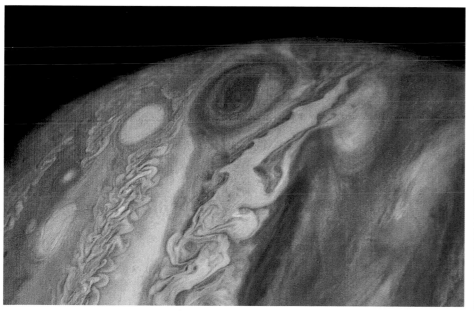

Fig 5 A massive storm of spiralling clouds three times the size of the Earth forms Jupiter's Great Red Spot.

Fig 8 Mars' surface is red due to iron oxide dust spread over its surface.

METEORITES

Fig 9 Interstellar diamonds. The Allende meteorite fell in Mexico in February 1969. Remaining unchanged since it formed, it is a chondrite, similar in composition to the primitive clumps of material from which the Earth formed. Under a transmission electron microscope (TEM) tiny diamonds can be seen, which are impregnated with gases captured from the atmospheres of stars that are even older than our Sun and Solar System.

Meteorites are natural objects which reach the Earth from great distances in space. Specimens can be collected in the field and, when studied, give us vital clues about the age, composition and internal structure of planetary bodies, including Earth, and about the origins of the Solar System (fig 9). Meteorites do not move fast enough in space to have escaped from the gravity fields of other stars, and consist of materials which are all found within the Solar System. The oldest meteorites are of similar age and include the oldest dated material in our Solar System. For these reasons, we regard meteorites as primitive debris left over from the Solar System's early history.

Detailed studies suggest that most meteorites are fragments of asteroids (page 5), pulled out of their usual orbit by Jupiter's gravity field, though some may come from comets orbiting within the inner Solar System. Just a few derive from the Moon and Mars, probably hurled into space as debris during previous huge meteorite impacts. When the orbits of meteorites cross the Earth's orbit, they may land on Earth. The smallest are dust-sized and the biggest have left spectacular craters up to 100 km in diameter. Apart from during the earliest stages of Earth history, however, such large meteorites have been exceedingly rare.

There are three main groups of meteorites: iron (fig 11), stony-iron (fig 10) and stony (figs 9,12). Iron meteorites consist of metallic iron mixed with a smaller amount of nickel, and many of them originated in the cores of their parent bodies, where they would have been completely molten. The Earth's core cannot be studied directly, but is thought to be similar in composition to iron meteorites.

Stony-iron meteorites are rare and probably come from a region above the core of their large parents. They are mixtures of iron-metal and silicate minerals, and represent the material that is likely to be present at the Earth's core–mantle boundary.

Most stony meteorites are chondrites. The rare remainder are achondrites, and these are basaltic, like the most common volcanic rocks of the Earth's crust and the outer layers of other stony planetary bodies. Their absolute ages (page 25) reveal when the outer layers of these bodies were last molten and first began to separate into their different layers. Chondrite composition varies but is broadly similar to that of the Sun without hydrogen and helium. Chondrites have remained unaltered since the beginning of the Solar System, whereas Earth rocks are still being continually changed by Earth processes (pages 10–11). Chondrites can also contain preserved grains of stardust – minute remains of other stars (fig 9). They therefore give glimpses of the raw materials of the Solar System, while their age tells us when it formed (around 4560 Ma ago).

Fig 10 Stony-iron meteorite from the Atacama Desert, Chile, 1823. Clusters of pale green olivine crystals are embedded in iron-nickel metal.

Fig 11 The cut, polished and etched surface of the Canyon Diablo meteorite reveals its high iron content.

Fig 12 A piece of the Barwell, (Leicestershire, England) stony meteorite which fell on Christmas Eve, 1965.

EARLY EARTH

The Earth continued to sweep up asteroids and comets in its path, but by a few million years after its formation it had reached its present size. Over the next 100 million years the planet began to differentiate into layers as gravity dragged the heaviest materials, such as the iron that forms about a third of Earth's mass, to the core. Lighter materials, such as iron and magnesium silicates, floated to the surface, creating the mantle and crust. The very lightest gaseous materials would eventually form a primitive atmosphere.

Throughout this time temperatures soared due to incessant pounding by meteorites and the release of internal heat from the decay of radioactive elements. By 4000 Ma ago the Earth's interior was probably producing more than five times as much heat as today, heating the planet to a few thousand degrees centigrade.

It was only after the Earth had settled into its 'onion skin' layers that the first crust could form (fig 13), but the question of exactly when the crust developed is vexing because most of the evidence has long since been absorbed in the Earth's highly efficient rock recycling factory. However, some evidence has recently been discovered in the form of microscopic crystals of zircon. These crystals have been dated up to 4400 Ma, making them the oldest mineral specimens on Earth. The presence of zircon grains proves that there were solid rocks at this time, which have since been eroded and recycled.

EARTH'S SATELLITE

The Moon is the Earth's sole natural satellite, although in recent years we have launched many small artificial satellites that now orbit the Earth, quenching our thirst for instant information by constantly relaying data across the globe. Compared to satellites around other planets, such as those of Jupiter, the Moon is big in relation to the size of the Earth.

It is thought that a massive planetesimal, up to three times the size of Mars, punctured the Earth's skin about 4500 Ma ago, generating almost enough heat to melt the planet. Part of the Earth's lightweight crust and mantle vaporized during the impact and combined with the colliding body to form our nearest neighbour. Using computer simulations, scientists have estimated that a solid ball the size of the Moon would have taken less than a year to form. The early Moon was probably closer to the Earth than it is at present, but ever since formation it has gradually spiralled further away, and continues to do so at a rate of some 3 cm each year.

Craters scarring the surface (fig 15) reveal that until about 3900 Ma ago meteorites must have pummelled both the Earth and the Moon. A meteoritic rain has continued to shower the Earth and Moon ever since, but fortunately a great deal less intensively and frequently – the majority being micrometeorites otherwise known as 'cosmic dust'.

Fig 13 Cross-section through the modern Earth (not to scale), showing iron core, heavier silicate mantle and lighter silicate crust. The core is nearly half the size of the Earth (12,760 km across) and the lithosphere is about 90 km thick. The asthenosphere behaves plastically and is a layer of partial melting. The size and density of the Earth's layers are worked out from the speed, behaviour and direction of earthquake vibrations (seismic waves) as they reach different places around the Earth.

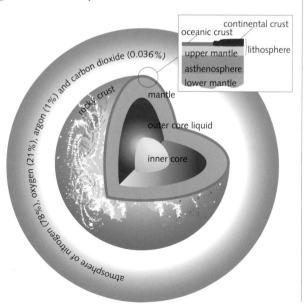

continental crust
oceanic crust
lithosphere
upper mantle
asthenosphere
lower mantle

carbon dioxide (0.036%)
argon (1%) and
oxygen (21%),
atmosphere of nitrogen (78%)

rocky crust
mantle
outer core liquid
inner core

Moon landings (*Apollo* missions) have added enormously to our knowledge of the Moon. It has a dusty 'soil' of debris, perhaps from meteorite impacts, and it has neither water nor an atmosphere. There is a silicate crust and a silicate mantle (fig 14). The maria ('seas') represent gigantic basaltic lava flows probably caused by partial melting of the mantle. These volcanic rocks however, are older than about 2700 Ma, and the Moon's surface, crust and upper mantle are now effectively dead, geologically. Although small moonquakes occur, rocks are not recycled in the Moon's upper layers, as they are on Earth (pages 10-11).

Fig 14 Moon cross-section (not to scale). The Moon is about a quarter the size of the Earth.

Fig 15 The pock-marked surface of the Moon with the Earth in the distance.

CRUST AND WATER

Fig 16 Molten Earth.

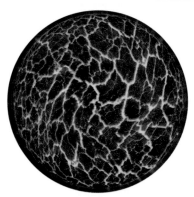

Fig 17 Cooling Earth.

Fig 18 Watery Earth.

Once the initial intense bombardment by meteorites subsided, the Earth's surface gradually began to cool, and, rather like a skin congealing on cooling custard, solid crust started to appear (figs 16, 17). It was probably similar to the brittle crust that forms when molten lava cools, but extended over huge areas.

The earliest evidence of this first crust is from komatiites (fig 19). These are volcanic rocks containing a high concentration of magnesium, indicating that they formed at temperatures of up to 1650°C, which is much hotter than any lavas erupting today. They would have cooled rapidly as lava erupted and spilled over the Earth's surface.

The oldest known rocks on Earth are around 4030 Ma old (fig 20). Although they have been altered since they were first produced, the structure of these rocks indicates that they formed from sediments that accumulated under water, proving that some shallow rivers or seas were already present on Earth.

By this time surface temperatures had dropped to less than 100°C, partly because, according to models of stellar evolution, the Sun was emitting about 25% less heat than today.

By 3500 Ma ago it is thought that temperatures had settled below the current maximum of +58°C. The cooling climate allowed steam from volcanoes to condense, creating global rainstorms that formed the early rivers, lakes and seas. Water soon flooded most of the surface, leaving just a few scattered volcanoes and isolated islands jutting out of the global ocean (fig 18).

Around this time the two types of Earth's crust also began to develop – continental crust and oceanic crust. Continental crust is light and forms a layer about 35 km thick. It is granitic, meaning that it is composed mainly of aluminium-rich silicates. Oceanic crust is dense and thin, forming a layer about 6 km thick. It is basaltic, meaning that it is composed mainly of iron and

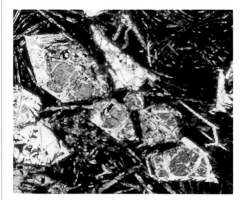

Fig 19 Texture of a komatiite (width 4 mm).

Fig 20 A specimen from the Amîtsoq Gneiss (3800 Ma old), western Greenland, close in age to the Earth's oldest-known rocks.

magnesium silicates. The formation of oceanic and continental crust also marked the start of the crustal cycle (fig 21) – a process in which the constant regeneration of oceanic crust, roughly balanced by its re-absorption into the mantle, breaks up and rearranges pieces of continental crust. This continual recycling means that no parts of the oceanic crust are more than about 200 Ma old, compared to the oldest areas of continental crust at over 4000 Ma.

EARTH'S CRUST

Rocks in the Earth's crust are formed in three different ways – igneous, metamorphic and sedimentary processes. **Igneous** rocks form when magma cools and solidifies. Basalt is a common igneous rock that forms when volcanoes spill magma (as lava) on to the Earth's surface. Granite is another common igneous rock that is produced when magma cools and solidifies slowly, deep within the crust. **Sedimentary** rocks are formed from the breakdown of pre-existing rocks by weathering and erosion, or they may be of organic or chemical origin. The sediments accumulate in layers and eventually create rocks such as mudstone, sandstone and limestone. **Metamorphic** rocks form when other rocks are altered by heat or pressure. Schist, gneiss, slate and marble are common metamorphic rocks.

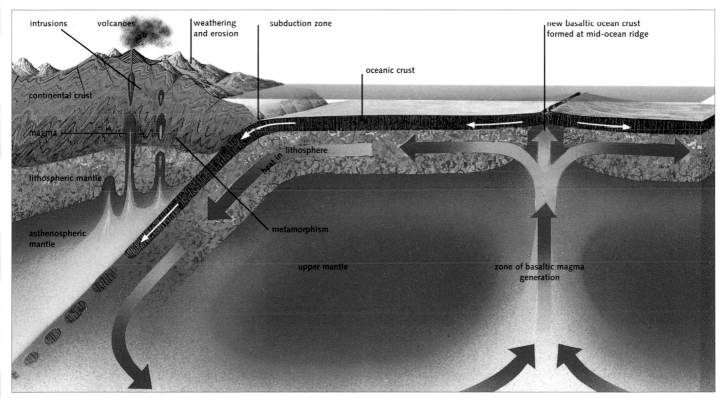

Fig 21 The Earth's crust is continually being recycled.

LIFE EVOLVES

By around 4000 Ma ago the Earth had permanent crust, seas, a virtually oxygen-free but carbon-dioxide-rich atmosphere, and temperatures within current ranges, providing a cradle in which life could evolve (fig 22). All living things contain many chemical elements, but the most essential ingredients for life are water, carbon, hydrogen, oxygen and nitrogen (which are available from compounds within the Earth), and energy (which can be released from the breakdown of chemical compounds or extracted from sunlight).

Microfossils 3500 Ma old are the earliest direct evidence of life (fig 23). These fossils reveal single cells held together to form long filaments that look like beads on a string. Their structure is similar to living species of cyanobacteria, which are single-celled photosynthetic organisms (page 15), some of which form the characteristic green carpet of scum on stagnant water. Their relatively complex cell structure indicates that the first signs of life would have evolved well before 3500 Ma ago, but, if so, the incessant baking and squeezing of the Earth's crust must have obliterated all imprints of such tiny, fragile life forms.

Because of the intense battering by meteorites, early life could not have survived on Earth before about 4000 Ma ago. However, scientists have discovered a chemical signature of life in rocks from Greenland that are 3850 Ma old. They contain a form of carbon found only in living things, suggesting that life

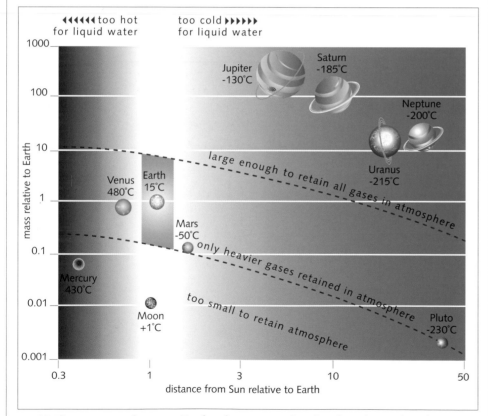

Fig 22 The unique conditions on Earth today, compared to the other planets, have been critical for the emergence and survival of life. (Planets not to scale.)

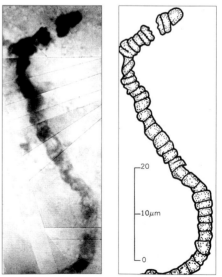

Fig 23 Electron micrograph and reconstruction of Primaevifilum amoenum, the earliest fossilized evidence of life from the Apex Chert of north-western Australia.

evolved as soon as conditions were stable enough for it to survive. Although we cannot pinpoint exactly when and where life originated, we know that, unlike today, early life forms evolved and thrived in an atmosphere lacking oxygen. Several theories about how life evolved have been proposed.

ORGANIC SOUP

A pool of molecules on the Earth's surface may have provided suitable ingredients and conditions for primitive life to evolve (fig 24). Charles Darwin first speculated about this idea of 'primeval soup' when he wrote in a letter in 1871, "It is often said that all the conditions for the first production of a living organism are now present, which could ever have been present. But if (and oh! what a big if!) we could conceive in some warm little pond, with all sorts of ammonia and phosphoric salts, light, heat, electricity, etc., present, that a proteine compound was chemically formed ready to undergo still more complex changes, at the present day such matter would be instantly devoured or absorbed, which would not have been the case before living creatures were formed."

The theory continued to have appeal through the 1950s and 1960s when Harold Urey and Stanley Miller re-created in the laboratory what were then believed to be the primitive conditions on Earth with an atmosphere of mainly methane, ammonia and hydrogen. On adding an energy source, the experiments succeeded in spontaneously generating complex molecules. However, because opinions on the composition of the Earth's early atmosphere have changed, these results are now generally discredited. It is currently thought that, early in Earth's history, carbon dioxide was present in high concentrations, conditions under which the experiments do not work. Also, regardless of the exact environmental conditions thought to be present when life evolved, a satisfactory explanation for how the molecules, once formed, could have assembled into self-maintaining, self-replicating life forms has not yet been put forward. Although this remains a problem, a clearer picture about the earliest forms of life is now emerging

Fig 24 Champagne Pool, Waiotapu Thermal Reserve, New Zealand, a hot volcanic spring rich in sulphides and bacteria.

HOT VENTS

More recent versions of Darwin's organic soup theory have focused on sulphur as the key to early life. The discovery in 1977 of black smokers – volcanic, deep-water, hydrothermal vents (fig 25) along mid-ocean ridges – led to the idea that early life evolved in such settings. Modern black smokers squirt out jets of scalding mineral soup rich in sulphur compounds, which support teeming bacteria. These provide food for other organisms in the thriving, specialised communities around these vents. The most primitive bacteria known today contain heat-shock compounds that help the production of proteins within the cell when temperatures are high, supporting the view that tolerance of high temperatures could be an ancient adaptation.

Fig 25 The Spire, a black smoker 18 m high on the Mid-Atlantic Ridge, 3080 m below sea level.

13

LIFE EVOLVES

Whereas photosynthetic organisms use light as their energy source, early vent organisms, like their modern counterparts, would have been chemosynthetic, unlocking the abundant sulphur compounds around them to obtain their energy. The energy is used, as in photosynthesis, to convert water and carbon dioxide into carbohydrate tissue. But modern bacteria around black smokers also take advantage of the oxygen dissolved in seawater as an additional energy source. This was not available to the earliest organisms, and, although the warmth around them would have speeded up their chemical reactions, they could not have lived close enough to the vents themselves to obtain sufficient heat to make up their energy gap.

Various kinds of hydrothermal hot springs (fumaroles) and pools are also found in volcanic regions at the Earth's surface (fig 26), some of them rich in sulphides and bacteria, just as Darwin might have wished for his "warm little ponds". In such shallow water environments, early life forms could have harnessed sunlight to make good their energy gap, using a more primitive form of photosynthesis which does not depend on oxygen. Primitive green and purple bacteria, existing in just this way, can be found today in hydrothermal ponds.

OUTER SPACE

It is clear that the Earth suffered an intense stream of collisions by asteroids and comets during its formation and early development, and that this may have influenced the timing of the evolution of life. But is it possible that life originated in outer space, and then became 'seeded' on Earth during an impact?

Meteorites do contain many of the ingredients necessary for life. Water is also required for life to evolve, but recent space missions have discovered that many of the other planets and their satellites have evidence of ice that may have been water at some stage. So, although there is no conclusive evidence

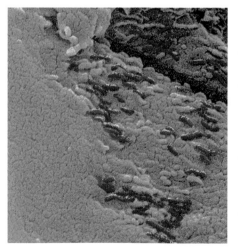

Fig 27 Coloured SEM of pink tube-like structures, possible microfossils in a meteorite from Mars (ALH 84001, Antartica, 1984). Structures are a fraction of the diameter of a human hair.

Fig 26 Hot springs; a geyser in Yellowstone National Park, USA.

yet, it is certainly possible, perhaps even inevitable, that life has evolved elsewhere in the Universe (fig 27).

The Earth today is unique among the planets in having an atmosphere containing 21% oxygen, but it was not always like this (fig 38). Rock and fossil evidence indicates that there was little or no free oxygen in the earliest atmosphere, and that over time the air we breathe has been modified by living things and their interaction with the environment.

The Earth's primitive atmosphere probably contained hydrogen, carbon monoxide, methane, nitrogen, ammonia and hydrogen sulphide. But by about 4200 Ma ago gases and steam ejected during volcanic eruptions – a process called outgassing – created an atmosphere of mainly carbon dioxide, nitrogen, sulphur dioxide and water vapour. There was barely any oxygen, but small amounts would have been produced when ultraviolet radiation from the Sun split water vapour into hydrogen and oxygen – a reaction called photodissociation. The newly-formed oxygen helped to create the ozone layer, which filtered out the ultraviolet rays, thus preventing further photodissociation and shielding the Earth from the harmful radiation. High levels of carbon dioxide in the early atmosphere produced a powerful greenhouse effect (fig 28), with heat from the Sun trapped by gases in the atmosphere.

All green plants today photosynthesize, which means that they capture energy from the Sun to convert carbon dioxide and water into carbohydrates, releasing oxygen as a waste product. Oxygen is an element that has profoundly altered our atmosphere, fundamentally affecting the course of biological evolution. From over 3000 Ma ago large volumes of it were being pumped out by early photosynthesizing organisms called cyanobacteria, which lived in watery environments. At first the oxygen would have dissolved into the surrounding waters, but by about 2000 Ma ago atmospheric levels rose significantly, once the waters were saturated.

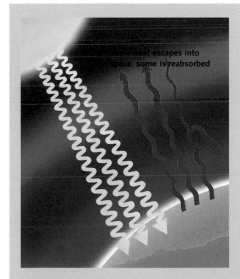

Fig 28 Schematic representation of the greenhouse effect.

THE GREENHOUSE EFFECT

Carbon dioxide in the atmosphere is essential to life on Earth. Along with other 'greenhouse gases' – such as water vapour, methane, CFCs, ozone and nitrous oxide – it helps balance the Earth's heat budget. Without these gases, the planet would be as inhospitable to life as the Moon.

Incoming solar radiation that is not reflected back into space by the upper layers of the atmosphere is absorbed by the rocks, waters and living organisms of the Earth's surface. In return, the warmed surface radiates energy out into space in the form of long-wave infra-red energy (heat). This energy is partially reabsorbed by the greenhouse gases in the air, keeping the Earth warmer than it would otherwise be by some 33°C. This process makes life on Earth possible.

The greenhouse effect is so called because it has been likened to the trapping of warmth by the glass in a greenhouse. The glass acts like the atmosphere by letting in light, but retaining heat. Global warming is thought to be happening because of human activities adding to the natural greenhouse effect. Predictions suggest that the world may be up to 5°C warmer a hundred years from now.

LIFE SHAPES ATMOSPHERE

Initially oxygen would have poisoned other organisms that were adapted to a reducing atmosphere – one with more methane and hydrogen than today. But gradually living organisms evolved that could use oxygen, which enabled them to produce more energy, become more complex, and in turn continue to modify the atmosphere.

Four types of evidence in the rock record have helped scientists to understand how and when oxygen began to form part of our atmosphere: fossilized cyanobacteria (stromatolites) that produced the first oxygen, and three kinds of rock (banded iron formations, red beds and palaeosols) that initially soaked up the oxygen. This oxygen history is a good example of how Earth and life have evolved together. This is the basis for James Lovelock's *Gaia* idea.

MICROBIAL MATS

Stromatolites (figs 29, 30) are formed by cyanobacteria, so fossil stromatolites prove that oxygen was being produced by photosynthesis at that time. They date from 3500 Ma ago, a time when these slow-growing communities had no grazing predators and therefore dominated the seashores. Today cyanobacteria are still common, but stromatolites rarely build up large structures because grazing invertebrates such as snails devour the newly formed microbial mats. However, some communities can still be found, for example, in remote bays along the coastlines of Western Australia and Florida in the USA.

Fig 29 Cross-section of stromatolite.

Fig 30 Stromatolite colonies living today in Little Darby Island, Exuma Cays, Bahamas.

ROCK RUST

Banded iron formations (or BIFs, fig 31) formed in shallow seas, mainly between 2500 and 2000 Ma ago, and in some areas of the world accumulated in layers hundreds of metres thick, providing the world's most important commercial source of iron. The red bands are composed of hematite, which is ferric (oxidized) iron, proving that oxygen was building up in the atmosphere and shallow seas during this time.

It is thought that BIFs formed when seawater seeped through cracks in the sea floor, collecting dissolved ferrous (reduced) iron from the Earth's mantle before being squirted back into the ocean through hydrothermal vents. The oxygen-starved deep water kept the iron in solution until ocean currents swept it into shallow continental shelf areas. The water here was newly oxygenated by photosynthesizing bacteria, which caused the iron to be oxidized and deposited as ferric iron grains in layers of sediments.

The alternate layers of iron-poor chert indicate that the oxidation process occurred in pulses, perhaps due to seasonal changes. The formation of BIFs would have soaked up much of the oxygen being produced up to 2000 Ma ago. However, their formation then tailed off abruptly and virtually ceased by 1700 Ma ago, probably because hydrothermal vent activity waned, reducing the supply of ferrous iron.

RED ROCK

Red beds (fig 32) formed in deserts, lakes or riverbeds rather than in the sea. As with BIFs, the rust colour is due to hematite, which formed when iron in the rock was oxidized by oxygen in the atmosphere. Rocks such as these have formed only during the last 2000 Ma. By this time enough oxygen had built up in the atmosphere to oxidize the land.

ANCIENT SOIL

Palaeosols (fig 33), which are the remains of ancient soils sandwiched between layers in the rock record, also reveal how much oxygen was present in the atmosphere at the time they formed. Palaeosols older than 2000 Ma contain ferrous iron, but slightly more recent examples contain large amounts of ferric iron, indicating that there was a sudden increase in atmospheric oxygen at that time.

Fig 31 Banded iron specimen, 7 cm thick and about 3000 Ma old, from the Murchison Goldfield, Western Australia.

Fig 32 Bright red sandstone coastal cliffs in Devon, England (from the Permian, around 260 Ma old).

Fig 33 Ferric palaeosols (dark) developed on volcanic pumice ash layers (pale), Tenerife.

17

SIMPLE TO COMPLEX CELLS

Bacteria and cyanobacteria are prokaryotes (fig 34a), which means that they are composed of small simple cells. For nearly 1500 Ma, from the first evidence of life around 3500 Ma ago, they were the only living organisms on Earth. Prokaryotic cells do not have a true nucleus, so their genetic material lies freely within the cell. This means that they usually reproduce asexually, creating a clone of the parent with identical genetic information.

All plants and animals, as well as some other single-celled organisms and fungi, are eukaryotes (fig 34b), which means that they are composed of larger complex cells. Organisms with eukaryotic cells developed at least 2000 Ma ago. The cells are more than 1000 times bigger than prokaryotic cells, a characteristic that has enabled the earliest fossilized eukaryotic cells to be identified as such.

Their advanced cell architecture includes a nucleus containing a large package of genetic material that enables eukaryotes to reproduce sexually. Each parent contributes a complete genetic blueprint; these are shuffled together to produce unique offspring with a new genetic code, which allows a potentially faster rate of evolutionary change to occur. In addition to sexual reproduction, eukaryotes have self-contained structures within the cell, called organelles, that carry out specific functions, such as respiration and photosynthesis.

JOINING FORCES

Eukaryotic cells probably evolved when several different types of free-living bacteria (prokaryotes) were captured by another bacterium. The theory, put forward by Lynn Margulis in the 1970s (fig 35), proposes that instead of digesting them, the cell played host to the captives, ultimately benefiting each party – a relationship called symbiosis. The merger would have bestowed novel metabolic capabilities on the new combined organism. The previously free-living bacteria would also have gained protection and a stable environment from settling within the host cell.

Although it is thought that eukaryotes started to evolve when oxygen levels in the atmosphere were very low, they would have been able to obtain oxygen from oxygen-bearing compounds such as silicates, nitrates and phosphates. Once oxygen levels rose, eukaryotes were already adapted to using oxygen so could immediately exploit this new energy source.

Several structures in eukaryotic cells point to an independent past. Green plant cells contain chloroplasts that enable plants to convert the Sun's energy into carbohydrates. Chloroplasts are thought to be descendants of free-living cyanobacteria. Mitochondria convert

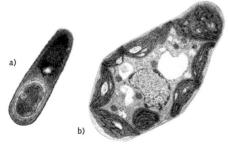

Fig 34 Electron micrographs (not to same scale) a) prokaryotic, and b) eukaryotic cells.

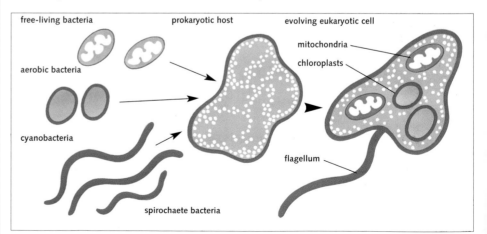

Fig 35 Complex cells may have evolved from a symbiotic relationship between simple cells.

oxygen to energy; a process called aerobic respiration. They can produce up to 19 times more energy than cells that use other means of energy production. They probably evolved from free-living aerobic bacteria. Flagella, which enable eukaryotes to move about, are similar in structure to a group of living bacteria called spirochaetes and probably evolved from them.

Multicellular organisms have a variety of cells that are specialized to perform specific tasks, such as digestion, movement and reproduction. The fossil *Mawsonites* (fig 36) is one of the earliest multicellular organisms. Similar to a jellyfish, it lived during the end of the Precambrian, between 549 and 543 Ma ago. It belongs to a group of early soft-bodied animals called the Ediacaran fauna, named after the South Australian locality where the fossilized impressions of these animals were first discovered.

Fig 36 Mawsonites spriggi, *from Flinders Range, Australia. Specimen 10 cm across.*

Fossils from this group of animals have now been unearthed elsewhere in the world, revealing a diversity of body architecture, ranging from the radial symmetry of *Mawsonites* to the bilateral symmetry of the quilted leaf-like *Dickensonia* (fig 37a). *Charnia* (fig 37b), named after Charnwood Forest in Leicestershire, England, where it was found, has a structure similar to a sea pen. It is still being debated among scientists whether these fossils represent the early prototypes for animal groups living today, or whether they were an evolutionary offshoot unrelated to present-day animals, but it is agreed that almost no trace of them survived into the Cambrian Period.

Around 545 Ma ago, at the beginning of the Cambrian Period, the number and variety of marine fossils suddenly increased. This is often called an 'explosion' of life, but it is more likely that animals became bigger and developed shells and skeletons at this time, hence they are more likely to become fossilized. Many small, soft-bodied animals must have already existed before the 'Cambrian explosion', but soft tissues rarely survive long enough to be preserved in the rocks, so scant evidence of them has been found.

It is this selective nature of fossilization that can sometimes create a misleading record of past life. For example, there are thousands of species of tiny marine crustaceans called copepods and they are fantastically abundant in today's oceans, yet only a handful of fossilized copepods have ever been discovered even though molecular research suggests that they probably evolved well over 200 Ma ago.

Many of the Cambrian fossils, for example the trilobites, are relatively complex animals that scientists believe could not have appeared suddenly 545 Ma ago, but must have had a long period of prior evolution. In support of this view, molecular studies indicate that basic examples of most animal groups may have started to evolve up to 1000 Ma ago. It is thought that most of these primitive animals were small and had no hard parts – perhaps being similar in form to the larval stages of marine organisms alive today.

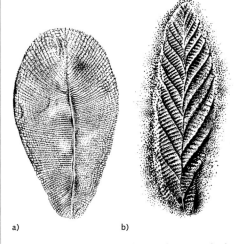

a)   b)

Fig 37 a) Dickensonia (*it is unknown which group this belongs to*), *b)* Charnia.

ATMOSPHERE SHAPES LIFE

OXYGEN

Oxygen has played a key role in the evolution of life. It reached its current atmospheric level of 21% by about 500–400 Ma ago, and although there have been minor fluctuations since then, it has remained fairly constant, peaking between 350 and 250 Ma ago (fig 38).

The level of oxygen in the atmosphere has remained so stable for the last 500–400 Ma because it is constantly being used and renewed by the interaction of Earth processes and living things. In fact, the living vegetation on Earth is so prolific in its production of oxygen, not only on land but also in the sea, that if it were not being removed at a constant rate, it would take only 5000–7000 years to accumulate the current volume. But as rapidly as it is being produced, oxygen is being soaked up by aerobic respiration, and during chemical weathering, soil formation and oxidation of volcanic gases.

The continuous recycling of oxygen keeps the volume at a constant rate, but how has it reached and remained about a fifth of our atmosphere? This question is currently being researched and is thought to involve a complex control mechanism, probably linked to the availability of phosphates and nitrates in the oceans.

CARBON DIOXIDE

While the oxygen level in the atmosphere has increased steadily over geological time, the level of carbon dioxide has fallen dramatically. It now forms just 0.03% of our atmosphere, but during the Archean the amount of carbon dioxide may have been many hundreds of times greater. This blanket of carbon dioxide would have created a strong greenhouse effect, maintaining temperatures between 0 and 15°C and preventing a global freeze despite the much weaker heat radiating from the Sun at this time. Even as recently as 550 Ma ago the amount of carbon dioxide in the atmosphere was 5–20 times the current level. The churning up of deep ocean waters, bringing carbon dioxide to the surface then into the atmosphere, may have caused the surge in carbon dioxide during the Permian.

At the end of the Carboniferous Period, large volumes of carbon dioxide were removed from the atmosphere by photosynthetic organisms, and subsequently stored as organic carbon in fossil fuels and carbonates. Over time the gradual decline in carbon dioxide is due to a steady transfer from the atmosphere to rocks, either as carbonate carbon (from shells and limestones) or as organic carbon (from the fleshy tissues of living organisms). Around 95% of the world's carbon supply is currently stored in rocks such as limestones, coal, oil and peat.

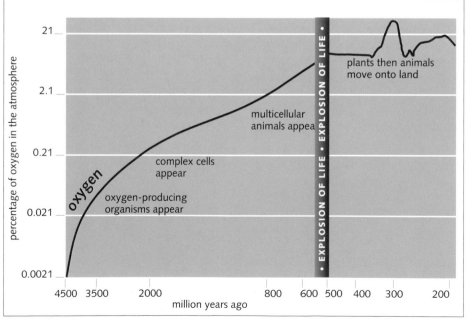

Fig 38 Graph showing the increase of oxygen in the atmosphere over time.

CHANGING OXYGEN LEVELS

Some of the main events resulting from changing oxygen levels are listed below.

Complex cells appear (2000 Ma)

Following the oxidation of ferrous iron on the Earth's surface (page 17), free oxygen built up in the atmosphere. This was a key factor in the evolution of eukaryotic cells (pages 18-19).

Size, shells and skeletons (550–495 Ma)

The combination of increasing oxygen in the atmosphere (fig 38), chemical changes in the sea, and the need for more support and protection from predators, probably triggered the development of shells and skeletons

The first shelly fossils were microscopic, e.g. Tommotian fauna around 540 Ma. Larger animals need high levels of oxygen in the atmosphere and oceans so that it can diffuse deep into their tissues. Oxygen levels appeared to reach such a critical threshold around 550 Ma ago (e.g. Ediacaran fauna; page 19). Further size and complexity followed with the development of circulatory systems to deliver oxygen around the body.

Land plants (410 Ma) and land animals (375 Ma) appear

It has been suggested that oxygen and corresponding ozone levels were sufficiently high at this point, to protect the plants and animals emerging from the water on to dry land (pages 34–35) from harmful ultraviolet radiation.

Waterproof eggs (350–250 Ma)

Reptiles were the first four-legged animals able to reproduce away from water. It is likely that high oxygen levels facilitated the evolution of amniotic eggs (page 38), which cushion the embryo in a watertight capsule while allowing oxygen and other gases to diffuse through the eggshell. High oxygen levels may also have favoured the evolution of giant aquatic invertebrates.

Fire limits (350–250 Ma)

Combustion of wood requires a 15% oxygen level in the atmosphere, and wood will combust spontaneously at 35%. The fossil record of charcoal and forest vegetation, and the occurrence of large fossilized tree trunks (fig 39), suggests that though forest fires (fig 40) occurred over this time, forests thrived. Oxygen levels must therefore have remained within these two limits.

Flying high (350–250 Ma)

Large-scale increase in vegetation leads to an increase in atmospheric oxygen through photosynthesis. It has been proposed that high oxygen levels may have contributed to the evolution of powered flapping flight – a high energy activity – as seen in flying insects, and later, in flying reptiles and birds (pages 46–47).

Fig 39 Fossilized tree trunks in cliff face.

Fig 40 Bush fire, Turkey Creek, Australia.

TIME IN THE ROCKS

NAMING THE LAYERS

The geological column divides time into two eons, subdivided into eras, which are further subdivided into periods. Rocks that represent a particular geological period are referred to as a system (e.g. rocks from the Cambrian Period form the Cambrian System).

Quaternary *1.8 Ma–now)* – fourth of the eras originally used to divide the last 545 Ma.

Tertiary *(65–1.8 Ma)* – third of four eras originally used to divide the last 545 Ma.

Cenozoic Era *(65 Ma–now)* – recent life.

Cretaceous *(142–65 Ma)* – after the Latin word *creta*, meaning chalk, which was abundant at this time.

Jurassic *(206–142 Ma)* – after the Jura Mountains in France and Switzerland.

Triassic *(248–206 Ma)* – after a characteristic triple sequence of rocks in Germany.

Mesozoic Era *(248–65 Ma)* – middle life; the second of the four eras originally used to divide the last 545 Ma of Earth's past.

Permian *(290–248 Ma)* – after the ancient Asiatic kingdom of Permia.

Carboniferous *(354–290 Ma)* – from the large amounts of organic carbon, which later turned into coal, deposited during this time.

Devonian *(417–354 Ma)* – after rock sequences in Devon, south-west England.

Silurian *(443–417 Ma)* – after *Silures*, a Celtic tribe living in southern Wales.

Ordovician *(495–443 Ma)* – after *Ordovices*, a Celtic tribe living in northern Wales.

Cambrian *(545–495 Ma)* – after the Roman word for Wales, *Cambria*; rocks of this age were first discovered in northern Wales.

Paleozoic Era *(545–248 Ma)* – ancient life; the primary or first of the four eras originally used to divide the last 545 Ma of Earth's past.

Phanerozoic Eon *(545 Ma-now)* – the time of visible life.

Proterozoic Era *(2500–545 Ma)* – the time after which organisms with complex cells (eukaryotes) evolved.

Archean Era *(3800–2500 Ma)* – the period when simple single-celled organisms (prokaryotes) were the only life on Earth.

Hadean Era *(4560–3800 Ma)* – the time span before life evolved.

Precambrian/Cryptozoic Eon *(4560–545 Ma)* – the time of hidden life and represents about 90% of Earth's past.

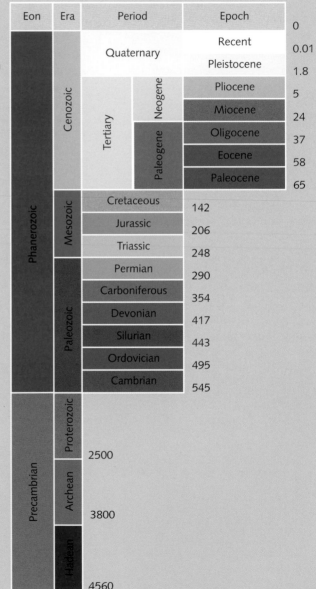

Eon	Era	Period		Epoch	
Phanerozoic	Cenozoic	Quaternary		Recent	0 / 0.01
				Pleistocene	1.8
		Tertiary	Neogene	Pliocene	5
				Miocene	24
			Paleogene	Oligocene	37
				Eocene	58
				Paleocene	65
	Mesozoic	Cretaceous			142
		Jurassic			206
		Triassic			248
	Paleozoic	Permian			290
		Carboniferous			354
		Devonian			417
		Silurian			443
		Ordovician			495
		Cambrian			545
Precambrian	Proterozoic				2500
	Archean				3800
	Hadean				4560

Fig 41 The geological column (figures are in million years).

Geologists examine the rocks beneath our feet and observe present-day Earth processes to interpret the past. To find out how old the rocks are, when events happened and how long they took, they study what the rock is made of and how and where it formed.

Fragments produced by the weathering and erosion of rocks are eventually deposited in layers of sediments, such as mud and sand. Over millions of years they are compressed into layers of sedimentary rock. As new layers are added over time a stack is formed with the oldest rocks at the bottom, the youngest at the top (fig 43) – this is known as the law of superposition.

Not only do the rock layers contain a record of the past, but their position in the sequence enables geologists to work out the chronological order of past events. Different parts of the Earth's crust have retained sequences of rock layers straddling different chunks of time. Geologists have pieced together these disparate sequences to create a single sequence spanning all 4560 Ma of geological time, called the geological column (page 22).

Sometimes there is a break in a sequence of rock layers where sediments have not been deposited or have been eroded and/or tilted, causing younger rocks to take their place above rocks that are much older – this is called an unconformity (figs 42, 44).

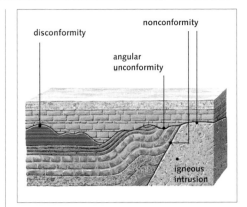

Fig 44 Unconformities fall into three main categories.

Fig 42 Angular unconformity, Gebel Esh Mellaha, Egypt, a break of 560 Ma.

Fig 43 The 2 km gash through the rocks at the Grand Canyon, USA, reveals nearly 3000 Ma of geological history.

DATING THE PAST

Fig 45 *Species of the sea urchin* Micraster *are important zone fossils from the Late Cretaceous (about natural size).*

Sedimentary rocks form as successive layers of sediments accumulate over time. But dating the layers is not always easy because:

- a single complete sequence of rock layers does not exist, as Earth processes have eroded, deformed and overturned different layers in different places;
- many different types of rock form in different places at the same time;
- the same rock type may occur at different times.

MATCHING FOSSILS AND ROCKS
Fossils enable scientists to unravel the rock layers. By studying the fossils found in successive layers – from older to younger – scientists have discovered the evolutionary sequence of life. This is because evolution proceeds in one direction only; it is essentially a non-repeated sequence of events. So each time interval has a unique array of fossils, which can be used to date the rocks in which they occur. If they contain the same fossils, rocks from various locations and depths in the Earth's crust are likely to be the same age. This method of estimating the age of rocks is called relative dating (fig 47). Relative ages tell us whether one rock is older or younger than another, but not their actual age in numbers of years (i.e. absolute age). Relative dating can also be used for rock sequences which do not contain fossils, by comparing other features such as their chemistry, minerals or sedimentation patterns, but these are not always practical especially across large areas of the Earth.

Some plant and animal groups, such as ammonites, are particularly useful for dating rocks because they have distinctive species that occurred over a short time span and had a wide geographical range. 'Fossil clocks' such as these are called zone fossils (fig 45). Microscopic organisms such as foraminifera are frequently used as zone fossils to date rocks, because they are abundant, and easily extracted and identified (fig 46). However, not even the best zone fossils occur worldwide. Marine organisms for example do not

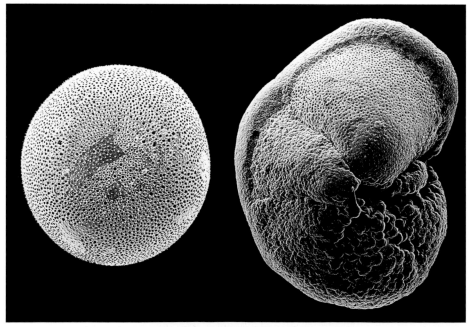

Fig 46 *Species of these foraminifera (left,* Orbilina: *right,* Globorotalia) *are used as zone fossils.*

24

occur in land deposits. Moreover, distinct biogeographical regions each have their own characteristic organisms (pages 54–55). It is common therefore, to have several different parallel dating schemes, which can be brought together when shared zone fossils are discovered, or by also using absolute dates.

RADIOACTIVE ROCKS

To find out exactly how old rocks are we need to look inside them. All rocks contain minerals made of elements. Some elements can exist in different forms, or isotopes, some of which are unstable and decay radioactively. If the mineral has not been changed since it formed, these unstable isotopes can act as precise geological clocks.

The clock starts from when the mineral crystallizes in the rock. As the rock ages, the radioactive isotopes decay at a steady rate and change into more stable isotopes. The steady rate at which they decay is measured by a time span called a half-life, which is the time it takes for half of the radioactive isotope in the rock to decay and turn into a more stable isotope. By comparing the proportion of radioactive isotope that remains in a rock to the proportion of stable isotope, we can calculate the rock's age in years – a method known as absolute dating. Uranium/lead (U/Pb), potassium/argon (K/Ar), fission track and carbon-14 (fig 48) dating are all forms of absolute dating based on radioactive decay.

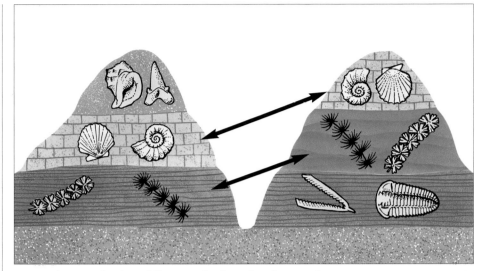

Fig 47 The same fossils in different rocks show that they are the same age.

Fig 48 This fragment of a human upper jaw from Kent's Cavern, Devon, England, has been directly dated to about 31,000 years old by radiocarbon dating. This shows that it is one of the earliest remains of modern humans in Europe.

WHAT ARE FOSSILS?

Fossils are evidence of living organisms preserved in rock for many millions or thousands of years, and give us a rare glimpse into the past. Only a tiny proportion of all the species that ever lived become fossils. This is because when a plant or animal dies it usually breaks up and decomposes or is scavenged, so particular conditions are required for it to fossilize and even then it is usually only the hard parts of an organism that are preserved. The fossil record of Earth's past becomes substantial from about 545 Ma ago up to the present day. Before this time, although life was well established, most organisms were small and soft-bodied so little fossil evidence of them survives.

Fossils can form in many different ways, but typically an animal dies and becomes buried in sediments and over time the hard parts leave impressions and/or are gradually replaced with minerals (fig 49). Once formed, fossils do not always remain intact – many are broken, distorted or destroyed if the rocks in which they are embedded are squeezed, heated or twisted – and they may need careful interpretation to work out the original form of the animal. Organisms may become fossilized close to where they lived. Others have been carried far away by winds or currents before becoming fossils.

TYPES OF FOSSILS

Replacements, moulds and replicas

Most fossils are formed by minerals replacing the original organic matter of the plant or animal (fig 50). When organisms are buried, sediments often solidify around them forming an external mould. Even if the original specimen there dissolves away, the mould remains in the surrounding rock. Sometimes the hollow cavities inside shells become filled with sediment. If the shell then dissolves away, an internal mould is left behind. In some cases the space between the internal and external mould is filled with minerals and becomes a replica – a natural cast of the original shell.

Fig 50 Ammonites are common replacement fossils; Asteroceras *(large),* Promicroceras *(small), Somerset, England.*

Trace fossils

Sometimes fossils record evidence of an organism's activity rather than the physical presence of the organism itself. Animal tracks, burrows and borings may be fossilized and are called trace fossils (fig 63).

a) A fish dies, falls to the bottom of the sea and becomes buried in sediments. The soft body parts rot away leaving the skeleton.

b) Sediment layers accumulate around the skeleton.

c) The skeleton is gradually replaced by minerals while the sediment becomes rock.

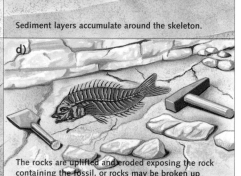

d) The rocks are uplifted and eroded exposing the rock containing the fossil, or rocks may be broken up revealing embedded fossils.

Fig 51 Soft-bodied animals such as jellyfish (Rhizostomites) may leave fossilized impressions (from Solnhofen, Germany).

Impressions

Plant leaves are often preserved as carbonized impressions (page 35). Skin, feathers and soft parts may leave an impression in the surrounding sediments (fig 51), but colour tends not to survive.

Wood to stone

Wood may be preserved when silica replaces the woody tissue, retaining fine detail such as growth rings and even cell structure (fig 52). This process is called petrification or petrefaction.

Fig 52 Fossilized wood of Psaronius, a tree fern.

Amber

Amber is fossilized tree resin that sometimes contains the remains of living organisms, mainly insects and spiders, which have been trapped as the resin hardened (fig 53).

Unchanged fossils

Not all fossils are formed of rock – some, such as teeth, may be largely made up of their original materials. There are some remarkable examples of whole mammoths trapped and frozen in Arctic tundra for thousands of years, and sabre-tooth cats and sloths drowned in La Brea tar pits in California, USA (reconstruction, inside back cover).

Fake and false fossils

Sometimes unusual rocks, such as nodules, are mistaken for fossils. There are also several well-known examples where fake fossils have been created to dupe scientists; for example, a forged fossil skull from England, called Piltdown Man (fig 54), was initially hailed as the 'missing link' between apes and humans.

Living fossils

Lingula is a brachiopod with a long trailing cord that anchors it to a vertical burrow in the tidal zone. Species belonging to this group evolved over 400 Ma ago and some have changed very little over time, so they are known today as 'living fossils' (fig 55).

Fig 53 Fungus gnat in Baltic amber.

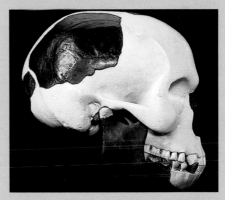

Fig 54 A model reconstruction of the famous Piltdown Man forgery.

Fig 55 Lingula, a living fossil, Indonesia (about 18 cm long).

RECONSTRUCTING THE PAST

At some time or other, nearly everyone has imagined having a very special vehicle that would enable them to travel back in time, as in H.G. Wells' famous book, *The Time Machine*. We can dismiss this as a pure science fiction fantasy, yet, in their own way, Earth scientists, like archaeologists and historians, have found effective methods of visiting the past.

Most of the information about the Earth's past, even for very large-scale events such as the changing geographical patterns of continents, seas and oceans, comes from slow and painstaking accumulation of small pieces of information, such as the record

Fig 56 Sir Charles Lyell.

of one particular fossil, or the chemical composition of one mineral.

The collection, recording and analysis of rocks and what they contain, provides raw information. In order to translate this information into a meaningful reconstruction of the past, we need models to interpret it. We understand the past by using a whole collection of interlocking models that Earth scientists are continually changing in the light of further research.

In order to create models of Earth processes, Earth scientists observe the Earth as it is today. We can study how rivers carry and deposit sediment, how volcanoes erupt, how desert winds make sand dunes, and what happens to the bones and shells of animals after they have died. In the simplest approach, we assume that these processes must have gone on in the past too.

The idea that we can use modern observations to make models of the past is called uniformitarianism, and is attributed to the Scottish geologist Sir Charles Lyell (1797–1875) (fig 56). But this certainly does not mean that the Earth's processes have never changed, rather that we assume that they have not changed until we find reason to think otherwise.

For instance, as we have already seen, much of the information held in the Earth's oldest rocks suggests that free oxygen was almost absent in the early seas and atmosphere, and that it increased over time (pages 20–21). This

is also consistent with progressive changes in life witnessed by the fossil record, and by evolutionary relationships among primitive organisms. Oxygen is so important to the present Earth that we have to picture the very early Earth and its processes as completely different from today. This does not mean that uniformitarianism is irrelevant. We just have to use it at a level which is relevant, in this case by studying those modern processes that take place in an oxygen-free environment.

There are essentially three kinds of information that are important in making models of the past. These answer the questions 'what?', 'when?' and 'where?' (figs 57, 58).

WHAT?

The most basic of all geological information is the composition of rocks, their chemical and mineral make-up, and the objects and features they contain, such as fossils. A fossil fish suggests that the deposit in which it was found was laid down under water. A rock consisting largely of natural glass suggests that it was once very hot and then cooled suddenly, as in a volcanic eruption.

Textures (e.g. coarse, fine) tell us about rates of cooling, or the strength of the currents that carried a sediment along. Fabrics (e.g. laminated, layered) suggest successive episodes of movement or deposition. Individual rocks are also grouped in characteristic ways within the Earth's crust, and are often tilted, folded and displaced along large rupture lines (faults). These can be read as indications of upheavals and subsidences, compression, heating, cooling and stretching.

WHEN?

No amount of description of the features of the Earth's crust alone will enable the past to be reconstructed in a coherent way. Earth scientists also need to relate everything to time. This might be relative and local, such as the way in which one fault crosses and displaces another and so must be the younger one. Where the folds of rocks have been truncated (cut across) by other rocks above, folding and erosion must have happened before the rocks above were deposited. For the global view, however, Earth scientists have also developed an international timescale which enables them to put their observations and interpretations into a worldwide sequence, the geological column (page 22).

WHERE?

Earth science is four dimensional. In addition to knowing how old everything is (the time dimension), we need to know how everything fits together spatially (i.e. in three dimensions). Rocks we see at the surface in one place disappear beneath younger rocks in other places or were once continuous with those elsewhere, but have been eroded. Every rock represents an environment of some kind, either within the crust or at the Earth's surface. We need to know the places where all the rocks of one particular age occur and what environments they represent so that we can map Earth's past.

Although their modern geographical position is relevant, it is also important to map the same set of rocks and environments in relation to ancient reconstructions of the continental positions. For example, in the early to middle Miocene, oxygen isotopes (page 58) indicate a period of global warming. This fits perfectly with records of warm-water corals occurring in central Europe, Tasmania and New Zealand, which were all much further from the Equator than any modern warm-water locality today.

Other clues about the Earth's past come from studying the present Earth, atmosphere and life forms, and also from astronomy and meteorites. Molecular analysis of living organisms gives us evolutionary histories, which can be compared with the fossil record. Even events that happen within our lifetime, such as an earthquake or a forest fire, can further our understanding of the ancient past.

Fig 57 Wadi landscape in the Oman Mountains. These rocks are part of an eroded slice of Cretaceous oceanic crust and upper mantle, thrust up on to the Arabian continent.

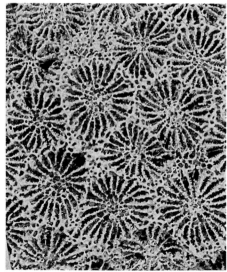

Fig 58 Perfectly preserved skeleton of a tropical coral (Goniastrea), Sussex, from the Eocene, indicating a warmer climate in southern England than today – also confirmed by its isotopic composition (page 58).

PERFECTLY PRESERVED

Fig 59 A prawn preserved in the Solnhofen Limestone, Germany.

On several occasions, when a sequence of chance events has conspired to preserve exquisite details of animals that would not normally be fossilized, we have a rare window on ancient life. Such fossils usually form when the plants and animals are rapidly buried in an oxygen-starved environment so their flesh does not decompose.

One example of such rare preservation comes from Solnhofen, which provides a snapshot of Jurassic life preserved in fine-grained limestone in Germany. The most famous fossil from this site is *Archaeopteryx*, the earliest known bird (fig 93), but dragonflies (fig 91), prawns (fig 59) and mammals have also been imprinted in the rock. The animals probably became entrapped in sticky mud when a lagoon drained at low tide.

However, perhaps the most famous example of rarely preserved fossils is the Burgess Shale in Canada, discovered early in the 20th century. These fossils reveal that many spectacular animals lived over 500 Ma ago during the Cambrian and are now extinct. They were entombed in mud and flattened over time into shiny films sandwiched between layers of black shale, preserving intact remarkable details of both the hard and soft parts of a variety of creatures.

The animals, mainly arthropods, worms and sponges, originally lived on a muddy sea floor in shallow waters. A mudslide probably swept them away from a sponge reef into deeper waters, where they were buried alive. Being surrounded by oxygen-starved sediments meant that rather than decaying, the soft body parts were preserved and eventually fossilized. To date, about 70,000 fossils have been collected from this remarkable site, providing a wealth of information about the diversity of Cambrian body designs.

THREE IN ONE

The largest known predator of Cambrian seas was an arthropod, called *Anomalocaris*, which reached up to 50–100 cm long. The story of reconstructing *Anomalocaris* is even more extraordinary than the giant 'prawn's' appearance, involving piecing together three different types of fossil fragments from the Burgess Shale (fig 60).

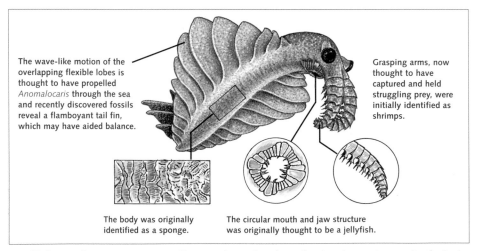

The wave-like motion of the overlapping flexible lobes is thought to have propelled *Anomalocaris* through the sea and recently discovered fossils reveal a flamboyant tail fin, which may have aided balance.

Grasping arms, now thought to have captured and held struggling prey, were initially identified as shrimps.

The body was originally identified as a sponge.

The circular mouth and jaw structure was originally thought to be a jellyfish.

Fig 60 Anomalocaris *was eventually pieced together from fossilized remains, each of which was originally thought to be a separate animal.*

WHICH WAY UP?

The bizarre creature *Hallucigenia* was named for its dream-like appearance. When scientists first studied its fossilized remains from the Burgess Shale they concluded that it must have walked on stilt-like legs and had a row of soft tentacles on its back waving through the water in search of food (fig 61a). It was thought to be unlike any animal living today.

However, the later discovery of other similar fossilized remains from China literally turned this interpretation upside down because a second row of tentacles was preserved. It is now agreed that the animal walked on paired tentacle-like limbs and used the spines along its back as protection from predators (fig 61b). But there is still no agreement on which end is the head and which the tail.

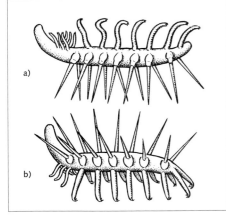

Fig 61 Older (a) and more recent (b) reconstructions of Hallucigenia.

TRILOBITE LIFE

Trilobites were the most common invertebrates during the Cambrian; their fossilized remains have been found in the Burgess Shale and all over the world (figs 63, 64). They are arthropods, which means that they have a hard external covering, a segmented body divided into head, thorax and abdomen, and jointed legs. They were the first animals to see the world around them, some possessing large compound eyes that gave them a wide field of vision. For nearly 300 Ma many species thrived in the oceans and seas, before their eventual demise 248 Ma ago during the mass extinction at the end of the Permian Period. Their hard carapace (shell), which they shed as they grew (fig 62), has meant that many trilobites have fossilized, enabling scientists to understand a great deal about their body structure and lifestyle.

Fig 63 Some trilobites were active swimmers using their legs to paddle through the water; others crawled along the sea-floor leaving distinctive trails.

Fig 64 Many trilobites (Calymene, West Midlands, England) rolled themselves into a protective ball when threatened, rather like woodlice today.

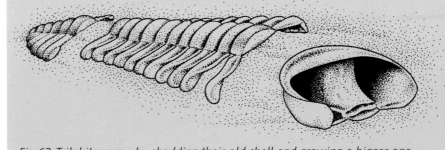

Fig 62 Trilobites grew by shedding their old shell and growing a bigger one.

THE LOST OCEAN

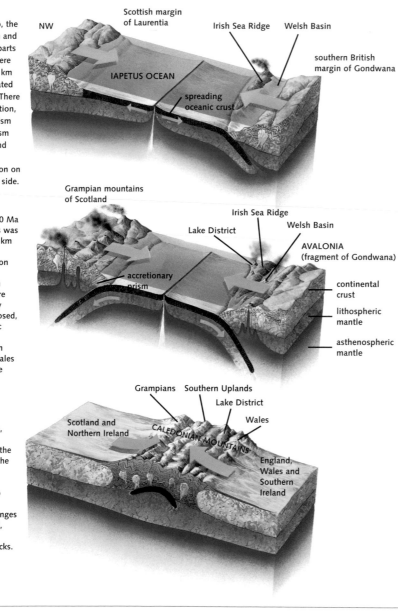

a) Around 550 Ma ago, the Gondwanan and Laurentian parts of Britain were about 5000 km apart separated by Iapetus. There was subduction, metamorphism and volcanism in Wales, and extensive sedimentation on the Scottish side.

NW
Scottish margin of Laurentia
Irish Sea Ridge
Welsh Basin
IAPETUS OCEAN
southern British margin of Gondwana
spreading oceanic crust

b) About 450 Ma ago, Iapetus was about 3000 km wide, with subduction on both sides. The Scottish margins were being highly metamorphosed, and volcanic island arcs developed in Scotland, Wales and the Lake District.

Grampian mountains of Scotland
Irish Sea Ridge
Lake District
Welsh Basin
AVALONIA (fragment of Gondwana)
accretionary prism
continental crust
lithospheric mantle
asthenospheric mantle

c) Around 420 Ma ago, Iapetus had closed, and the margins of the colliding landmasses buckled into a series of mountain ranges of contorted, cooked and squeezed rocks.

Grampians
Southern Uplands
Lake District
Wales
Scotland and Northern Ireland
CALEDONIAN MOUNTAINS
England, Wales and Southern Ireland

Fig 65 The closure of the Iapetus Ocean.

Over 500 Ma ago an ocean, called the Iapetus Ocean, divided Britain in two (fig 66). At that time the Lake District and Scotland would have been roughly the same distance apart as Britain and the USA today. The northern parts of what is now Britain were part of the same landmass as North America. The Iapetus Ocean began to form by sea-floor spreading at least 700 Ma ago. It probably reached about 5000 km in width, separating Laurentia (including areas of land that now form Scotland, North America and Greenland) from Gondwana (including areas of land that now form southern Britain, Europe, Africa, South America, India, Australia and Antarctica).

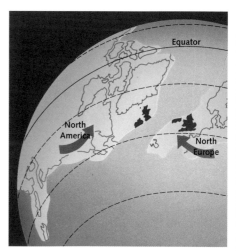

Fig 66 About 440 Ma ago, Avalonia (including southern Britain), had joined Baltica to form northern Europe which was approaching Laurentia (North America with Scotland).

Over millions of years the ocean floor beneath Iapetus was subducted, pushed under the continental landmasses, and the ocean began to close, disappearing completely by 400 Ma ago when the continents collided (fig 65). Britain was then joined, creating a seam or suture running across the British Isles from southern Scotland and the Solway Firth to the Shannon Estuary. The land buckling caused by the collision created mountain ranges in Europe and north-east America. The opening and closure of this ocean is just one example of the continuous cycle of sea-floor spreading and subduction that has been occurring on Earth since the formation of oceanic crust.

Several types of evidence reveal the existence of the Iapetus Ocean:
• When Iapetus was closing the ocean crust was subducted under the continental crust, but some stray pieces got pushed up on to the continental crust – these are called ophiolites.
• Rocks found near the land-join reveal that while Iapetus was closing a series of volcanic centres formed above the subduction zone. As one crustal plate was drawn down beneath another their movements heated up the crust. Vast reservoirs of molten rock (magma) formed, which spread deep within the crust to form granites and erupted violently at the surface of the overriding plate (figs 67, 68).

• Sedimentary rocks now lying across the land-join indicate that there was a range of different marine environments, e.g. beaches, shorelines and shallow shelves on either side, with deep ocean basin deposits between them.
• Shallow water fossils discovered in Scotland and North America differ from those found in the Lake District, Wales and the rest of Europe. The species found in Scotland and North America typically lived in warm-water environments in tropical regions, but those from England and Wales would have thrived in cold waters further from the Equator. This indicates that the fossils must have been buried in very different latitudes and a great distance apart.

Fig 67 *The volcanic rocks of the Lake District, England formed as Iapetus was closing.*

Fig 68 *Glowing ash and larger glassy fragments were blown violently from an island arc volcano. They formed a hot, dense cloud that moved rapidly down the flanks of the volcano and fused together as they settled, forming this rock, an ignimbrite, collected from foreground of fig 67.*

33

LIFE MOVES ON TO LAND

Before land vegetation evolved, the Earth's surface would have been barren and parched, with sweeping winds whipping up sand and dust storms. With no protective cover the land would have been baked and chilled on a daily basis. Because the surface was unable to soak up the erratic rainfall, flash flooding would have been common. Fossilized spores indicate that the first land plants evolved during the Middle Ordovician, but the earliest fossils of the plants themselves are 410 Ma old (fig 69). Remaining near the water and hugging the ground to take advantage of the layer of still, moist air, these primitive plants were small with a simple leafless branching structure.

About 400 Ma ago, soon after land plants evolved, they developed vascular tissue in the form of two tubes; one enabled them to carry water and nutrients from the soil, the other to transport manufactured food through the plant. In addition to a plumbing system, land plants need to support themselves against gravity using woody structures made of lignin, which form a rigid framework. To anchor themselves to the ground and to withstand battering by the elements, land plants need strong roots, which are also used to extract water and nutrients from the soil.

A waxy coating or cuticle helps prevent drying out and protects plants against airborne diseases. Pores within the waxy layer allow plants to regulate water loss, gases entering and leaving the plant, and temperature. Living on land also exposes plants to more sunlight than in water, which could be harmful if ultraviolet radiation is too strong. The movement on to land may have coincided with the build up of sufficient oxygen and ozone in the atmosphere to filter out the most damaging ultraviolet radiation. But the plant's waxy coating may also filter harmful rays, and some pigments offer additional protection.

Relatively soon after plants had conquered the land most of the major plant groups evolved (figs 70–73), rapidly transforming the landscape by adding dead matter to the soil, anchoring it with their roots, breaking down rocks and forming a protective and diverse layer of vegetation above.

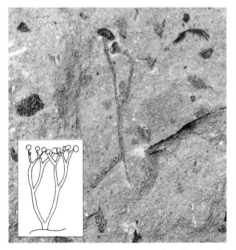

Fig 69 Cooksonia is an example of one of the earliest land plants, 417–391 Ma old.

34

PLANT DIVERSITY
Many of the main plant groups had evolved by the Carboniferous Period, although the flowering plants and a few others did not appear until much later.

Lycopsids (Late Silurian to Recent) were abundant during the Carboniferous, forming up to 50% of the plant fossils from this period. Their leaves were arranged in spirals around the trunk and branches, and left a regular pattern of scars when they were shed. *Lepidodendron*, the largest lycopsid reached heights of up to 40 m. The shallow rooting parts of lycopsids called *Stigmaria*, had radiating rootlets submerged in the swampy ground. Today, only small herbaceous relatives of the lycopsids survive.

Fig 70 Fossil leaf scars on stem of Lepidodendron.

Horsetails (Devonian to Recent), such as *Calamites*, reached up to 20 m in height during the Carboniferous, but only small relatives are alive today. They had hollow segmented trunks, which often filled with sediments when they died, producing a natural cast of the internal structure called a pith cast.

True ferns (Devonian to Recent) are ancestors of ferns living today. They reproduced using spores.

Fig 72 Frond of Sphenopteris laiventi, *Derbyshire, England (frond is 9 cm long).*

Seed ferns or pteridosperms (Carboniferous to Permian) are now extinct. They looked similar to the true ferns but reproduced using seeds.

Conifers (Carboniferous to Recent) are a living group of seed plants that includes pines and redwood. Fossilized ancestors of Christmas trees about 310 Ma old have recently been discovered in County Durham, England. A tropical flood swept the trees into a swamp where they were rapidly petrified, retaining remarkable detail of their internal structure.

Bennettites (Triassic to Cretaceous) evolved later during the Triassic. They are now extinct but looked similar to modern-day cycads (Permian to Recent), which are evergreens with a squat trunk crowned with leafy fronds.

Angiosperms (Cretaceous to Recent), the flowering plants, did not appear until about 125 Ma ago but now constitute more than 80% of living land plants. They diversified rapidly during the Cretaceous and have dominated land vegetation ever since. Fossilized wood, leaves, fruits and more rarely flowers (fig 74), seeds and pollen have been discovered (page 48).

Fig 71 Base of horsetail stem, Calamites, *from Yorkshire, England, about 300 Ma old.*

Cordaites (Carboniferous to Permian) are now extinct, but include the ancestors of living conifers. Some trees reached up to 30 m in height.

Fig 73 Frond of Alethopteris lonchitica *from Yorkshire, England, about 300 Ma old.*

Fig 74 Fossil flower Porana *from Baden, Germany, about 10 Ma old.*

COAL FORESTS AND CARBON CYCLES

The first tropical rainforests developed around 320 Ma ago on low-lying land covered with swamps, lakes and rivers. Plants flourished in the warm climate, creating vast swamp forests covering a total area roughly the size of North America (5000 x 700 km).

Dead plants usually rot because they are exposed to the air, which enables decomposing organisms to oxidize the carbon, releasing carbon dioxide into the atmosphere. However, in the swamp forests the stagnant, acidic water prevented dead plants from rotting. Instead, root and stem fibres accumulated and matted together forming peat and trapping the carbon. Over millions of years as the layers of peat piled up, all the water was squeezed out and the plant material broken down into hydrocarbons, eventually producing the coal we burn today. About 7000 years of plant growth and deposition are needed to form a 1 m seam of coal.

The repeated drowning and burial of swamp forests meant that much carbon dioxide was taken from the atmosphere and locked underground. Large amounts of limestone deposited during the Late Carboniferous also removed carbon dioxide from the atmosphere.

As these two processes trapped more and more carbon underground, there was a steady drop in atmospheric carbon dioxide levels, which reduced the greenhouse effect and led to cooling. This, combined with landmasses moving over the South Pole, caused the Earth to became colder, eventually triggering another ice age. The swamp forests dried out, driving some of the plants to extinction.

The earliest footprints on land have been discovered in the Lake District, northern England. They belong to a millipede-like animal that stepped out of the water on to barren volcanic land around 450 Ma ago, well before the appearance of land plants. However, once plants evolved on land they created moist, shaded environments as well as food sources that had not existed before, and an abundance of millipedes and other arthropods rapidly emerged from the water to exploit the new resources, resulting in an astonishing

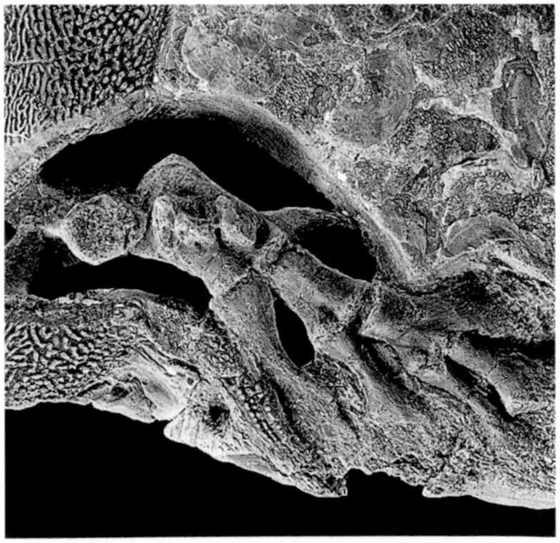

Fig 75 The jointed bone structure in the fin of Eusthenopteron *is equivalent to the skeleton of the human arm. The big bone at the left end corresponds to the humerus or upper arm bone (field of view about 8 cm).*

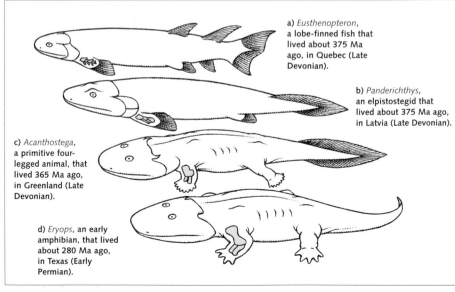

a) *Eusthenopteron*, a lobe-finned fish that lived about 375 Ma ago, in Quebec (Late Devonian).

b) *Panderichthys*, an elpistostegid that lived about 375 Ma ago, in Latvia (Late Devonian).

c) *Acanthostega*, a primitive four-legged animal, that lived 365 Ma ago, in Greenland (Late Devonian).

d) *Eryops*, an early amphibian, that lived about 280 Ma ago, in Texas (Early Permian).

Fig 76 Amphibians evolved from lobe-finned fish.

diversity of insects by the Carboniferous. Fishes remained in the water for some time after plants moved on to land but one group adapted to life in the shallows at the water's edge, and eventually moved onto land (fig 76).

Some lobe-finned fishes evolved lungs, enabling them to breathe air if the water grew stagnant – the living lungfishes are surviving examples. They also developed jointed bones inside their fleshy lobed fins (figs 75, 76a), which probably helped in swimming. When land plants and land arthropods were already well established about 380 Ma ago, some lobe-finned fishes ventured on to land (fig 76b). However with flattened bodies and eyes on top of their head, they probably lived and hunted mainly in very shallow water.

In these primitive four-legged animals paired fins evolved into limbs, which had fingers and toes but were still quite weak (fig 76c). The animals could walk on land but still had tailfins and gills and probably spent most of their time in water. They may have used their limbs to walk through shallow weed-choked water. Gills and tailfins were eventually lost as these four-legged animals spent more and more time on land and evolved into the first amphibians and reptiles (figs 76d, 77). They levered their bodies off the ground using elbow and knee joints.

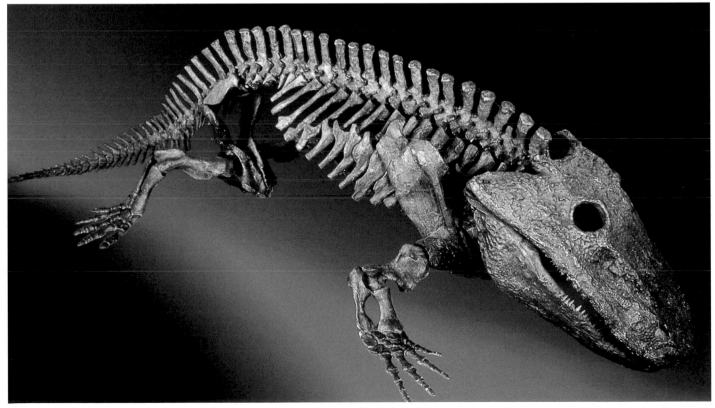

Fig 77 Eryops (fig 76d) had a robust skeleton to support its weight on land, but it had to return to the water to breed (about 1.5 m long).

37

AMPHIBIANS AND REPTILES

During the Carboniferous, around 330 Ma ago, early land animals split into two major groups – amphibians and reptiles. Amphibians lay their eggs (fig 78a) in water and breathe partly through their moist skin surface, which means that they can live in air or water but must return to water, sometimes migrating long distances, to breed.

Reptiles do not need to live near water. They have waterproof skins and lay eggs on land. These waterproof capsules, protected by a shell and containing the egg bathed in fluid, are called amniotic eggs (figs 78b, 79). The egg provides all the nourishment the embryo needs to develop until it

hatches and is able to survive in air. Because reptiles invest so much energy in each egg, they lay fewer of them, but each has a much higher chance of survival than the vast numbers of eggs spawned by amphibians.

The first amphibians and reptiles evolved in warm, humid, tropical regions, which had a constant climate and food supply. The earliest reptiles were small and similar in appearance to lizards today, but they soon began to adapt to drier environments.

Throughout the Late Carboniferous and Permian the dry climate produced large areas of arid and desert environments, which the reptiles were

well suited to, and they soon became the dominant large animals.

Although reptiles have adapted to a range of climates, and there has even been speculation that some dinosaurs were warm-blooded, all modern reptiles are cold-blooded, which means that their body temperature varies according to their level of activity and the outside temperature. In contrast, warm-blooded animals, such as birds and mammals, maintain a constant body temperature irrespective of external temperature. *Dimetrodon* (fig 81) was a common Permian reptile with a spectacular sail back supported by elongated spines from the vertebrae. It is thought that the

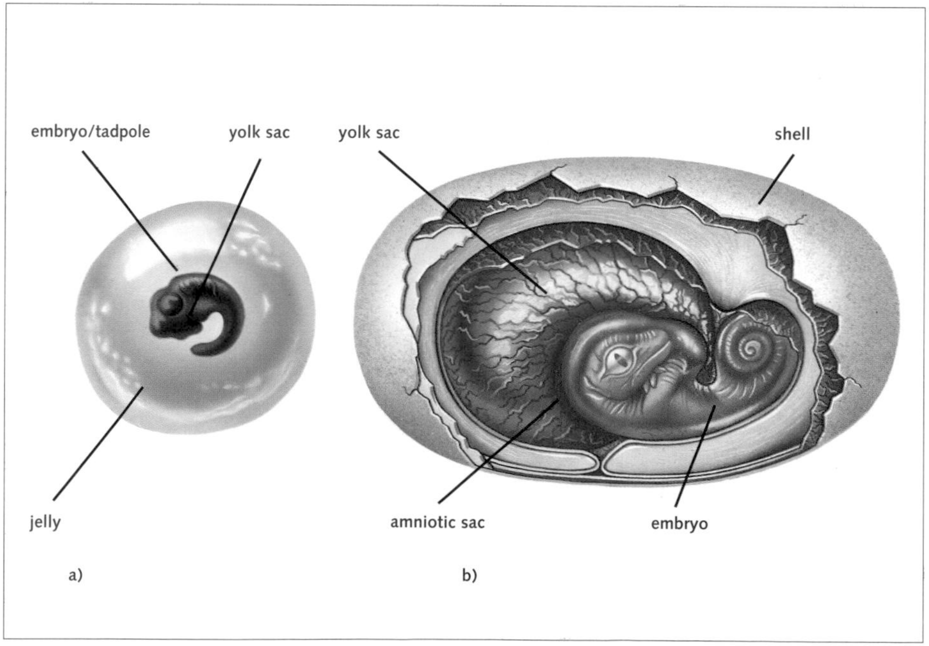

Fig 79 These turtle eggs (Testudo, Cirencester, England) are over 165 Ma old and are some of the oldest fossils of amniotic eggs (about 4 cm long).

embryo/tadpole   yolk sac   yolk sac   shell

jelly

a)

amniotic sac   embryo

b)

Fig 78 An amphibian egg (a) and an amniotic egg (b).

sail acted rather like a solar panel, absorbing heat from the Sun, which was then distributed by the blood circulation.

Although many reptile groups were severely reduced by the Permian extinction (page 40), the group, including the dinosaurs, continued to dominate the land for the next 150 Ma. One group of reptiles, the mammal-like reptiles, also gave rise to the mammals. *Dimetrodon* belonged to this group.

REPTILE CHARACTERISTICS
What characteristics do reptiles have to enable them to live on land?
- a skeleton for support;
- flexible limb joints to help them move;
- a shelled egg with a waterproof membrane, enabling them to reproduce away from water;
- a waterproof skin to prevent drying out;
- lungs to breathe air.

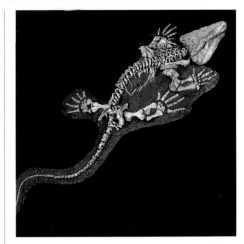

Fig 82 Limnoscelis is a 'proto-reptile' that did not lay amniotic eggs so had to return to the water to reproduce (New Mexico, USA, about 2 m long).

80 These fossilized footprints are thought to have been made by an amphibian, Megapezia, around 340 Ma ago (Yorkshire, England). The largest footprint is about 7 cm across.

Fig 81 Reconstruction of Dimetrodon, Early Permian, about 3 m long.

39

BOOM AND BUST

The overall diversity of life has escalated over time, despite several major setbacks. In total some 1000–3000 million species are estimated to have evolved, yet only 12.5 million currently exist on Earth today, which means that the majority of species that ever lived are now extinct. The evolution of new species is often roughly balanced by other species becoming extinct. But at least five mass extinctions have occurred during Earth's past (fig 83). These are triggered when an environmental crisis drives many species to extinction worldwide over a relatively short time on the geological timescale – usually less than several million years. After a mass extinction the groups that survive tend to diversify greatly, although it may take tens of millions of years to regain the original level of diversity.

The most severe Phanerozoic mass extinction occurred 248 Ma ago when about 90% of marine species became extinct, especially shallow water ones such as sea scorpions and trilobites. Many theories have been proposed,

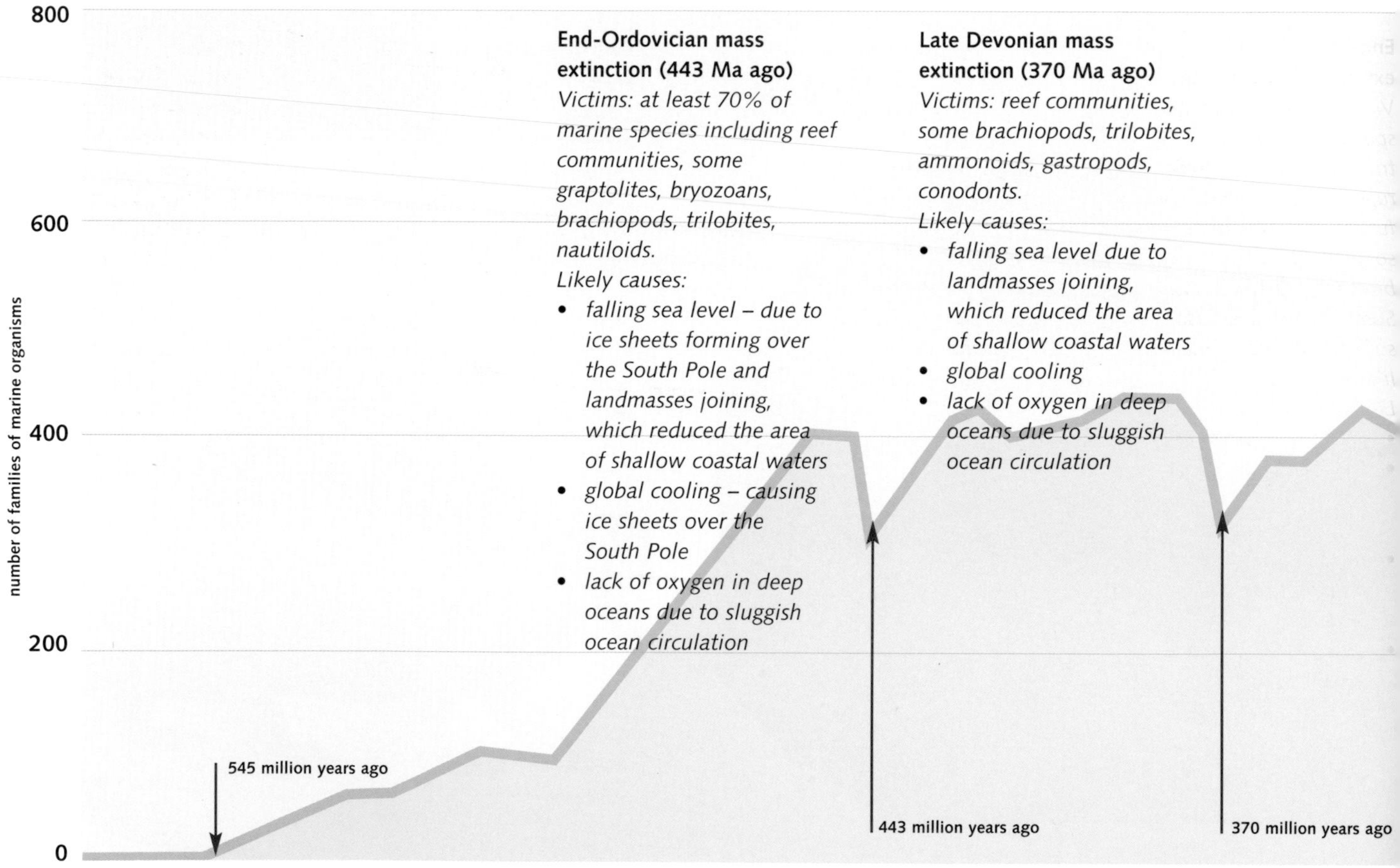

800
600
400
200
0
number of families of marine organisms
545 million years ago
443 million years ago
370 million years ago
End-Ordovician mass extinction (443 Ma ago)
Victims: at least 70% of marine species including reef communities, some graptolites, bryozoans, brachiopods, trilobites, nautiloids.
Likely causes:
• falling sea level – due to ice sheets forming over the South Pole and landmasses joining, which reduced the area of shallow coastal waters
• global cooling – causing ice sheets over the South Pole
• lack of oxygen in deep oceans due to sluggish ocean circulation
Late Devonian mass extinction (370 Ma ago)
Victims: reef communities, some brachiopods, trilobites, ammonoids, gastropods, conodonts.
Likely causes:
• falling sea level due to landmasses joining, which reduced the area of shallow coastal waters
• global cooling
• lack of oxygen in deep oceans due to sluggish ocean circulation

Fig 83 Biodiversity and mass extinctions over time. The likely causes of each mass extinction may have worked together or, in some cases, may be alternatives (pages 42–45). (Time axis not to scale.)

but a combination of relatively sudden global environmental changes is generally thought to have caused the mass death, in particular climate warming and oxygen shortage in the oceans. The weathering of Permo-Carboniferous coals during mountain-building and intensive volcanic eruptions at around the same time released large amounts of carbon dioxide into the atmosphere, causing global temperature to rise (page 44). This frenetic volcanic activity produced the Siberian Traps, massive layered rock steps formed by lava repeatedly flooding across the surface. The repetitive nature of the eruptions would have caused the climate to oscillate wildly over too short a time span for species to adapt to environmental changes, driving them to extinction. At the same time, the warmer climate would have reduced the solubility of oxygen in water and caused ocean circulation to become sluggish, leading to oxygen-starved oceans.

End-Permian mass extinction (248 Ma ago)
Victims: 90% of marine species including all trilobites, sea scorpions, rugosan and tabulate corals, fusulinid foraminifera, and some brachiopods, bryozoans, ammonoids. Some land species also suffered, including mammal-like reptiles.
Likely causes:
- *global warming*
- *lack of oxygen in deep oceans due to sluggish ocean circulation*
- *massive volcanic eruptions in what is now Siberia*
- *possibly falling sea level – when the supercontinent of Pangaea formed, the area of shallow coastal waters was reduced*

Late Triassic mass extinction (210 Ma ago)
Victims: all conodonts, some ammonoids, brachiopods, gastropods, bivalves.
Likely causes:
- *falling sea level*
- *massive volcanic eruptions along the eastern coast of North America*

End-Cretaceous mass extinction (65 Ma ago)
Victims: dinosaurs, plesiosaurs, mosasaurs, flying reptiles, ammonites, rudists and belemnites.
Likely causes:
- *fluctuating sea level*
- *asteroid impact*
- *massive volcanic eruptions in India and Pacific Ocean*

Sixth mass extinction? (now)
Humans have hunted animals for food for hundreds of thousands of years and have more recently destroyed large areas of natural habitats, driving many species to extinction. If this scale of destruction continues some scientists estimate that up to half of today's species could be lost in the next few hundred years.

248 million years ago

210 million years ago

65 million years ago

now

WHAT CAUSES MASS EXTINCTIONS?

Much research has focussed on what caused the five big mass extinctions, leading to an abundance of theories invoking both terrestrial and extra-terrestrial factors. The five events that have been given most serious consideration are volcanic eruptions, fluctuating sea level and climate change (fig 84); also asteroid and comet impacts, and reduced oceanic oxygen.

However, no matter what mechanism initiated each extinction, it is generally agreed that it is the resulting habitat destruction that causes the species loss rather than the event itself. Recently scientists have concluded that mass extinctions are more likely to occur when several events coincide or produce combined effects rather than as a result of a single catastrophe. In addition, the events contributing to mass extinctions often operate on vastly different timescales, from gradual changes of climate to the rapid effects of an impact.

IMPACTS FROM OUTER SPACE

Asteroids or comets colliding with the Earth may have caused or influenced mass extinctions. Some scientists have even suggested that regular strikes have caused extinctions every 26 Ma for the last 250 Ma, a cycle related to an increase in the number of comets passing close to the Earth, triggered by the passage of the Sun through the galactic plane.

Even though this mechanism is a matter of debate, there is undisputed evidence of a large impact 65 Ma ago that coincided with the mass extinction of the dinosaurs and other animals. An impact of this scale would produce sudden heat followed by a sharp fall in temperature as clouds of dust block sunlight. A rise in temperature would follow, caused by greenhouse gases released into the atmosphere from the sudden increase in the rate of death and decay of plants and animals.

VOLCANIC ERUPTIONS

Intense but relatively brief volcanic phases have coincided with at least two mass extinctions – the end-Permian 248 Ma ago when the Siberian Traps formed, and the end-Cretaceous 65 Ma ago when the Deccan Traps formed (figs 86). These large-scale eruptions would have engulfed the Earth in clouds of volcanic dust (fig 85), causing global darkness by preventing sunlight reaching the surface. Of the noxious gases belching into the atmosphere, sulphur dioxide produced acid rain, and carbon dioxide caused rapid global warming.

FLUCTUATING SEA LEVEL

Sea level has fluctuated widely over the last 550 Ma with rapid falls then rises

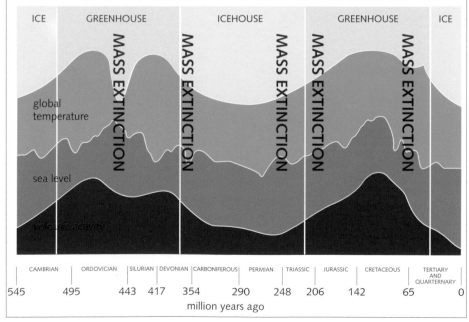

	CAMBRIAN	ORDOVICIAN	SILURIAN	DEVONIAN	CARBONIFEROUS	PERMIAN	TRIASSIC	JURASSIC	CRETACEOUS	TERTIARY AND QUARTERNARY
545		495		443	417	354	290	248	206	142	65	0

million years ago

Fig 84 Changes in volcanic activity, climate and sea level influence each other and may cause mass extinctions.

coinciding with most of the big five mass extinctions (fig 84). When sea level changes, species in shallow marine habitats and low-lying land areas are affected as their habitats are lost or radically altered. Sea level often rises again rapidly after a significant drop, occasionally flooding the continental shelf with deeper ocean water. If these deeper waters globally lack oxygen, this can be a major cause of marine extinctions. In the geological record, evidence of oxygen-starved seas is revealed in the formation of black shales. High sea level also floods areas of low-lying land, displacing land species.

Fig 85 Eruption of Soufrière, Montserrat, West Indies, December 1997.

Fig 86 The layered step structure of the Deccan Traps formed as successive lakes of lava erupted over the area (now deeply eroded), Martory Point, western India.

WHAT CAUSES MASS EXTINCTIONS?

There are several causes of sea level change, but one of the main influences is climate. Sea level rises when the Earth is warm because polar ice melts, but it falls when the Earth is cold because water is locked up as ice around the Poles and in mountainous regions. Seafloor spreading also affects sea level, which rises when the rate of spreading increases because the mid-ocean ridges become bigger and displace ocean waters upwards, so flooding the shelves. Movement of continental landmasses may also change sea level because, when landmasses collide, the overall land area shrinks and there is a corresponding increase in ocean area, reducing sea level. In addition there may be less coastline and therefore a reduced area of shallow water habitats.

CLIMATE PENDULUM

Climate is the long-term sum of weather patterns over time, and can be examined at both regional and global scales. During at least the last 650 Ma the Earth's climate has fluctuated between cold icehouse phases, when there were ice caps around the Poles, and warm greenhouse phases, when there was little or no polar ice (fig 84). What causes the climate pendulum to swing between warm and cold climates that last millions of years?

One model put forward to explain the fluctuations is that the Earth's dynamic interior drives climate cycles. The ripples from this enormous internal engine are

experienced on the Earth's surface in the form of gliding landmasses and erupting volcanoes. Volcanic activity, both on land and under water along mid-ocean ridges, releases carbon dioxide from the Earth's interior. It is the amount of carbon dioxide in the atmosphere that dictates climate. This is because of the role of carbon dioxide in the greenhouse effect (page 15).

It is estimated that the greenhouse effect raises average global temperature by 33°C and that without the protective blanket of gas the Earth would be completely covered with ice. When the level of carbon dioxide in the atmosphere is consistently high, the Earth's climate is warmer – a greenhouse phase. When the carbon dioxide level falls, the Earth's climate becomes colder and ice forms around the Poles – an icehouse phase (fig 88). The Earth is currently in an

icehouse phase, and for much of the last 545 Ma the Earth has been much warmer than at present.

Most wildlife is adapted to a specific climatic zone, such as tropical, temperate or polar habitats (fig 87). When global climate cools gradually, species tend to migrate towards the Equator, and similarly when climate warms gradually, species spread out towards the Poles. Environments and organisms interact with each other to form ecosystems that are always in a state of minor flux. But it is the sudden or drastic environmental changes, caused by falling sea level, asteroid or comet impacts, extensive volcanic eruptions and landmasses separating or joining, that are most detrimental to life on Earth because species, that do not have enough time to move to more favourable regions, are driven to extinction.

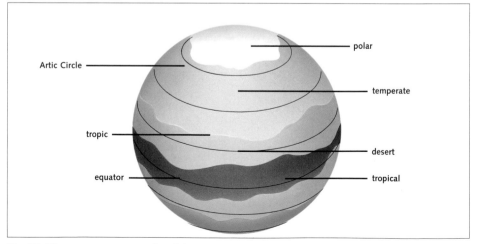

Fig 87 Climate zones across the globe.

WHAT CAUSES MASS EXTINCTIONS?

FREEZE AND THAW

Fig 88 Antarctic ice. The modern Earth is in an icehouse phase, with polar ice caps.

1. ICEHOUSE EARTH

The amount of volcanic activity fluctuates over time, and when volcanoes are less active, less carbon dioxide is released from the Earth's interior. But on the Earth's surface, carbon is constantly being buried in the form of carbonate rocks and preserved organic matter, causing the amount of carbon dioxide in the atmosphere to fall, and so the greenhouse effect is reduced. Climate cools and ice sheets form, causing sea level to drop.

4. FEEDBACK

The Earth does not overheat because eventually the amount of carbon dioxide in the atmosphere begins to fall. This occurs when the carbon in the organic matter generated by increased productivity is buried as carbonate rocks, composed of the shells and skeletons of marine organisms, and as preserved organic carbon, such as remains of land plants and phytoplankton. As the level of carbon dioxide drops, so climate cools and the Earth enters another icehouse phase.

2. FEEDBACK

The Earth does not continue to freeze because when sea level falls the newly exposed rocks, including limestones and other carbon-rich rocks, are then weathered, which releases carbon dioxide back into the atmosphere so warming the Earth.

3. GREENHOUSE EARTH

Increased volcanic activity, both on land and under water along mid-ocean ridges, releases large amounts of carbon dioxide into the atmosphere. When the amount of carbon dioxide in the atmosphere is high, the greenhouse effect increases, trapping more of the Sun's heat. Climate warms and sea level rises triggering a rise in organic productivity which is reflected in increased abundance and diversity of life.

Fig 89 Namib Desert, Sossulevei, Namibia. Deserts and other hot regions are more extensive during a greenhouse phase.

SOARING THE SKIES

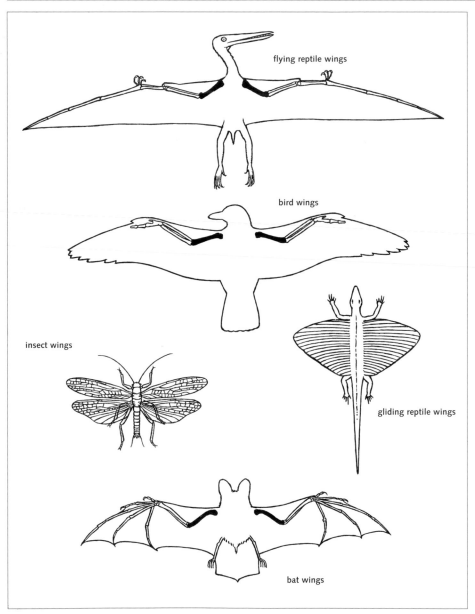

flying reptile wings

bird wings

insect wings

gliding reptile wings

bat wings

The Mesozoic Era witnessed a prolific increase in the number of animals exploiting the skies, as birds and several groups of reptiles independently acquired the ability to fly (fig 90). Mammals also eventually took to the skies, but not until the Tertiary. Some flying animals developed a large flat surface area to glide through the air; others had powerful flapping wings to propel themselves upwards and forwards. Most larger flying animals have a light, strong and streamlined body shape.

The first animals to fly were insects (fig 91), about 320 Ma ago back in the Carboniferous Period. Insect wings are

Fig 91 The delicate veins of a dragonfly wing (Cymatophlebia – wingspan 14 cm) are preserved in the Solnhofen Limestone, Germany (about 150 Ma old).

Fig 90 Comparison of wing structures. The wings of the flying reptile, bird and bat are homologous; they are supported by the bones of the forearm.

very thin delicate membranes composed mostly of dead tissue that cannot be repaired. In common with all arthropods, insects cannot grow bigger unless they shed their rigid outer skeleton and grow a new one. For flying insects this would include new wings, which is costly in terms of energy and vulnerability to attack during moulting. To avoid this potential risk, most adult flying insects do not moult, limiting their growth to the larval phase of development.

Small gliding reptiles evolved in the Late Permian, over 250 Ma ago. Elongated ribs with skin stretched across them formed wing flaps extending from their body rather like those of flying lizards today, which can glide up to 60 m between trees. In some early gliding reptiles the ribs were jointed so their wings could probably be folded against the body when not in use. Both insects and gliding reptiles retained their limbs as well as developing wings, so they could walk and climb unhindered.

However, the most abundant flying reptiles during the Mesozoic were the pterosaurs (fig 92). All members of this group were capable of powered, flapping flight, but larger species relied on gliding once in the air. They had lightweight skeletons with air spaces in their bones. To support the wing membrane pterosaurs had an extremely long, bony fourth finger and a sturdy breastbone to anchor the powerful flight muscles. *Pteranodon* was a large flying reptile that lived about 90 Ma ago, during the Late Cretaceous. Possessing wingspans of about 7 m, they probably had a similar lifestyle to today's albatrosses, skimming the sea in search of fish prey.

Birds are warm-blooded and are unique in having feathers. Unlike the pterosaurs, the fingers are reduced and support only part of the wing. The earliest known bird, *Archaeopteryx* (fig 93), is 147 Ma old and had flight feathers on its arms and a tail like a bird. However, it also had curving hand claws, a long, jointed tail and sharp teeth like a reptile, leading most scientists to conclude that birds descended from small meat-eating dinosaurs. It is still being hotly debated how the first birds became airborne – whether it was by running along the ground flapping their feathered arms, or launching themselves from the treetops and gliding through the air prior to evolving powered flapping flight.

Bats are flying mammals that had evolved by about 55 Ma ago. Their wings differ from those of pterosaurs and birds because they use four fingers as spokes to support the wing membrane, which is stretched between the arm, body and leg. Their walking is limited and they usually hang upside down when not flying.

Fig 92 This pterosaur, with a wingspan of 46 cm, chased and caught its prey on the wing (Pterodactylus) *(Solnhofen Limestone, Germany, about 155 Ma old).*

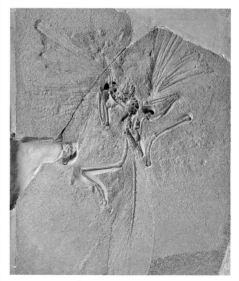

Fig 93 One of just seven exquisitely preserved fossils of Archaeopteryx *that have been discovered in the Solnhofen Limestone, Germany.*

47

INTO BLOOM

Fig 94 A fossil seed from a tropical plant, Anonaspermum, *common in England about 40 Ma ago.*

Fig 95 Leaf of an early flowering plant, Magnolia, *from Saxony, Germany (about 24 cm long).*

During the Cretaceous Period the flowering plants evolved and soon began to dominate the landscape. By the end of the period, magnolia, oak, beech, holly and willow had appeared and woodlands would have resembled those growing today.

Flowering plants are the most common of all living plants and include about 250,000 species. In addition to clothes and shelter, they provide most of our food, such as rice, cereal crops, beans, fruits, vegetables, coffee, tea, alcoholic drinks and spices. Also, most of our medicines were originally derived from plant extracts, such as aspirin from willow and the contraceptive pill from the yam. Yet they evolved relatively recently – around 125 Ma ago – in contrast to the ferns and conifers that evolved well over 300 Ma ago.

The key to the success of flowering plants is that they can reproduce in the driest of conditions by enclosing the developing embryo in a waterproof and nourishing seed, which also protects against infection and insect attack. In the absence of water they depend on wind or animals for the fertilization and dispersal of their seeds.

Many flowering plants produce brightly coloured and highly scented flowers with sweet energy-rich nectar, which attract animals, particularly insects, to visit frequently and thereby pollinate them. The evolution of flowering plants (figs 94, 95, 97) seems to have coincided with a massive diversification of insect groups as the moths, butterflies, ants, bees and wasps (fig 96) evolved and flourished.

Fig 96 Fossilized wing of a digger wasp (family Sphecidae) about 130 Ma old from Surrey, England. These wasps were one of the first groups of insects to pollinate flowering plants. Pollen would have stuck to their faces as they fed on it, and been transfered to other flowers the wasp visited.

Fig 97 Fossilized flowers are extremely rare. Although flowering plants evolved about 125 Ma ago, this fossil (Porana from Baden, Germany), is much more recent.

Fig 98 Riu Carreu gorge cuts through a Late Cretaceous carbonate platform at Vilanoveta in the Catalonian Pyrenees.

About 100 Ma ago, soon after the flowering plants emerged, the Earth was warm and the sea reached its highest ever level of around 250 m above present-day level. Large volumes of carbon dioxide were being pumped into the atmosphere by active seafloor spreading at mid-ocean ridges, and by volcanoes erupting on land and under water, particularly in the western Pacific.

As the atmospheric carbon dioxide level rose, the greenhouse effect increased and temperatures became an estimated 8–13°C higher than today. The difference in temperature between polar and tropical seas lessened when the Earth heated up, resulting in sluggish ocean circulation. Life in the warm languid seas bloomed, resulting in rapid carbon burial in the form of several types of sedimentation: extensive carbonate platforms in warmer seas (fig 98), widespread Chalk deposits in higher latitudes (fig 99), and oily reservoirs.

The shallow waters of the Tethys Ocean, which lay across the tropics, supported thriving communities of rudist bivalves, corals and other shelled animals. Over time their shelly remains slowly built up layer upon layer of sediment on the sea floor, on shelves and platforms on which further communities continued to flourish in the warm shallow water (fig 100). Around 93 Ma ago these carbonate platforms covered an area roughly the size of Europe.

The warmth and nutrient abundance also triggered an explosion in the abundance of marine plankton, producing a persistent rain of decaying organic matter on the shallow sea floor. Isolated pockets of stagnant warm water lacking oxygen meant that, in places, the organic remains were preserved and trapped in rock layers. Over time the remains turned into oil, some of which was slowly squeezed into porous sedimentary rocks nearby.

When sea level reached its peak, warm shallow oceans teeming with tiny planktonic algae called coccolithophores flooded mid-latitude landmasses such as Europe and North America. The coccolithophore remains gradually accumulated, forming the thick Chalk deposits seen in the striking cliffs of southern and eastern England (fig 99).

Fig 99 Chalk cliffs of Old Harry Rocks, near Swanage, England.

49

DINOSAUR DEATH

Fig 100 A large (about 15cm) rudist bivalve Vaccinites galloprovincialis *about 85 Ma old, from the late Cretaceous carbonate platform at Vilanoveta (see fig 98).*

Around 65 Ma ago 65% of species on Earth were wiped out – not just the dinosaurs (fig 101), but also pterosaurs, plesiosaurs, mosasaurs, ammonites (fig 102), belemnites, some bivalves including rudists (fig 100), several types of plankton and many other invertebrates all became extinct. Some species declined gradually during the preceding 10 Ma, while others, including many species of marine plankton, died out suddenly at the end of the Cretaceous.

The causes of this mass extinction have been hotly debated for many years. The focus of many theories is the global layer of clay laid down 65 Ma ago, about 2.5 cm thick, that has much higher levels of iridium than surrounding layers. Rarely found in the Earth's crust, iridium is common in the Earth's core and in extra-terrestrial bodies. Scientists have proposed that a vast extra-terrestrial impact or extensive volcanic eruptions could explain the unusual mineral layer.

It is now thought that three environmental crises coincided with the end-Cretaceous extinction – falling sea level, an extra-terrestrial impact and volcanic eruptions. We will consider each to assess whether a single event or combination of events triggered the mass death on land and in the sea. Around 65 Ma ago sea level fluctuated

Fig 101 The notorious Cretaceous Tyrannosaurus rex; *robotic reconstruction.*

Fig 102 Nostroceras, *Germany. Ammonites with unusual coiling were common in the Late Cretaceous.*

rapidly, at one point dropping almost 100 m below the present level. Climate changes and habitat destruction resulting from changing sea level would have contributed to the loss of some species.

Minute fractures in quartz crystals from around the Yucatan Peninsula of Mexico (fig 103), and submerged remains of a vast crater (fig 104), suggest that a large asteroid or comet, up to 18 km in diameter, collided with Earth about 65 Ma ago, but it is not clear what part the impact played in the mass extinction. An impact of this scale would have produced sudden heat followed by near freezing temperatures as clouds of dust blocked sunlight for about 6 months. A rise in temperature would have followed, caused by greenhouse gases released into the atmosphere.

However, the feature that sets this impact event apart from others is that it struck rocks rich in sulphur. The sudden addition of tonnes of sulphur aerosols into the atmosphere would have reduced light levels and prevented photosynthesis for several months, destroying the fragile web of interactions between plants and animals.

Extensive volcanic eruptions in India and the Pacific also occurred about 65 Ma ago. For over a million years vast sheets of molten lava pulsed across southern India, forming the Deccan Traps – a basalt plateau several kilometres thick (fig 86). Clouds of volcanic dust would have reduced sunlight. Greenhouse gases, such as carbon dioxide, belching into

the atmosphere would have caused rapid global warming. Each wave of volcanic activity would have meant that the climate was disturbed for several years before returning to normal, and most living organisms would have been unable to survive such erratic climate swings.

It is clear that the disasters outlined above all took place at the end of the Cretaceous, but no single event seems to be responsible for the escalating extinction rates. It is more likely that it was the catastrophic culmination of the three events, and the resulting devastation of habitats, that caused the mass extinction over several million years.

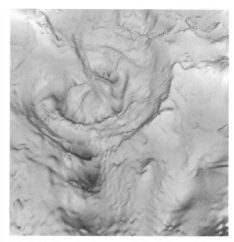

Fig 104 Gravity image of the Chicxulub crater, Yucatan peninsula, Mexico, 180 km across.

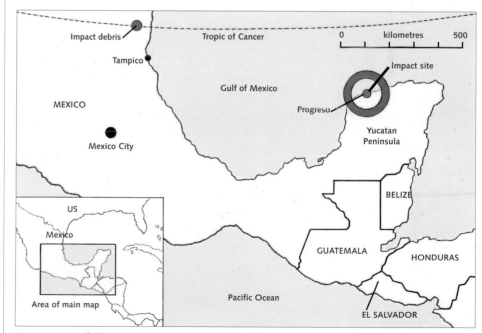

Fig 103 Site of Chicxulub crater centre.

MAMMALS DIVERSIFY

When the dinosaur dominance abruptly ended, a plethora of ecological habitats became vacant. However, this did not last long, as the tiny insect-eating mammals, previously confined to scuttling around at night, emerged into the daylight, became bigger and rapidly diversified to exploit habitats on land, in freshwater and the sea, and in the air.

Mammals are warm-blooded hairy animals. The fact that they also suckle their young and have complex teeth gives mammals two important advantages over their predecessors. The mother can feed the young without having to abandon them in search of food. In addition, the variety of tooth structure means that, across the group, mammals are equipped with the tools to shear, pierce, crush, grind or gnaw different foodstuffs; some species exclusively eat plants, insects or larger meat prey, while others have a more eclectic diet of animals and plants.

Mammals (fig 106) are well known as the group to which humans belong, but number far fewer species than birds, fishes and reptiles, not to mention the invertebrate groups. The ancestors of mammals are the mammal-like reptiles, (fig 81), which originated during the Late Carboniferous. From this group the first true mammals evolved about 225 Ma ago during the Triassic Period, but for about 150 million years, until the start of the Tertiary, they were small nocturnal mainly insect-eaters living in the shadow of the dinosaurs.

GRASSLANDS SPREAD

The Earth began to get considerably cooler from around 35 Ma ago. One of the effects of the cooling climate was that, by about 18 Ma ago, grasslands gradually started to replace large areas of forest. Prior to this time most herbivorous mammals were browsers eating soft flowers and leaves, but, as grasslands spread, animals such as horses evolved grazing habits.

Grasses are low-growing flowering plants. Grazing animals eat the top of plants but grasses survive because they grow from the base of the blade rather than the tip. Grass also contains tiny silica fragments – called phytoliths – making it abrasive. However, grazing mammals evolved teeth with high crowns (fig 105b), which in some species grow continuously, counteracting the increased wear from eating grass.

The cooling of Earth's climate which first became marked in Antarctica about 35 Ma ago culminated in our modern ice caps (page 57). The ever-shifting continents are probably one cause of this, particularly the fragmentation of Gondwana since the Mesozoic (fig 137). Northward movements of Gondwanan continents, broke up Tethyan seaways, and opened the Southern Ocean.

Fig 105 Lower cheek teeth of fossil horses:
a) low-crowned browsing type, Mesohippus, *about 33 Ma old; b) high-crowned grazing type,* Equus, *about 200,000 years old.*

a)   b)

Fig 106 A condylarth, an early Tertiary mammal.

Oceanic circulation became confined to separate oceans, as today, reducing the movement of warm water into higher latitudes, so polar regions cooled and ice caps grew (pages 56–57). During the Tertiary, collisions of Africa, Arabia and India with northern landmasses created the gigantic natural rampart of the Alpine–Himalayan fold-belt, whose position and sheer height restricted northward movement of warm moist air into higher latitudes, and intensified the Asian monsoon. Cooler, drier climates developed to the north, as in the Gobi Desert.

Why are the Himalaya (figs 108, 109) and Tibet so high? India collided with Asia around 45 Ma ago (fig 107), but because continental crust is too light to be subducted, India became wedged beneath the Asian margin. Instead of coming to a halt, it has continued to move 2500 km into Asia at 45 mm per year, thrusting and thickening the Asian continental crust of Tibet in front of it. This Tibetan thickening, and the double continental crust of the wedged zone (now the Himalaya), have buoyed them upwards. The scale and speed of uplift and weathering has significantly removed carbon dioxide from the atmosphere, reducing the greenhouse effect, so further contributing to long-term cooling.

Fig 108 The extraordinary height of the Himalaya (including Mount Everest at 8.8 km above sea level) has long fascinated both geologists and mountaineers.

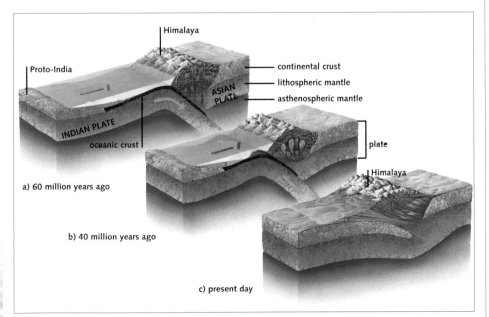

a) 60 million years ago

b) 40 million years ago

c) present day

Fig 107 Simplified model of the formation of the Himalaya showing subduction of the Indian plate.

Fig 109 When landmasses collide, large areas of the sea-floor are folded and uplifted. Marine fossils, like this ammonite (Virgatosphinctes, about 9 cm across, from over 5000 m in the Himalaya), may be raised to great heights far from their origin.

53

ON THE MOVE

It was over 200 Ma ago that the supercontinent of Pangaea began breaking into smaller land blocks (fig 137), and started isolating many plant and animal populations. The continuing shuffling of the continents since then has had a

Fig 110 Nine-banded armadillo (Dasypus novemcinctus) *living today in Texas, USA.*

profound effect on the past and current distribution and evolution of plants and animals (the study of which is called biogeography).

When drifting landmasses moved into different climatic zones, conditions became unsuitable for many of the plants and animals that already existed on and around them. Many species became extinct, but the isolation also caused new forms to emerge, which were better suited to the new conditions. In some cases, drifting continents later collided with one another forming new combinations of landmasses. This radically changed patterns of animal and plant distribution, through extinctions and intermixing.

Similar changes occurred in the seas and oceans as new seaways formed between the separating landmasses while others became closed off from each other. The richness of organisms in different places on Earth is often related to climatic warmth, and on land this is further affected by humidity. In general, the richest variety of life occurs where the climate is warmest and most humid, but the potential of a region can only be realized if new species emerge or if existing species can colonize the region from elsewhere.

ONCE AN ISLAND

South America had broken away from both Antarctica and Africa by 80 Ma ago and remained an island until relatively recently. During this time South American mammals and birds evolved in isolation and included:

- armadillos and their relatives, the glyptodonts, which had heavy body armour for protection against predators (fig 110);
- large ground sloths whose only relatives now live in trees;
- marsupial carnivores whitch nurtured their young in pouches, such as the wolf-like *Borhyaena* and the sabre-tooth *Thylacosmilus*;
- large flightless carnivorous birds with powerful tearing beaks and shredding claws (fig 111);
- many hoofed placental mammals – now extinct – which give birth to fully-developed young.

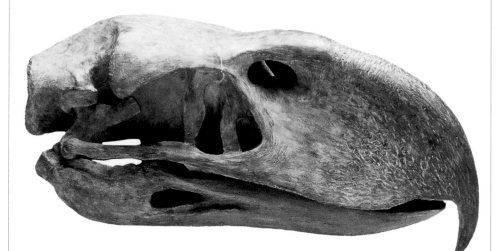

Fig 111 Large flightless birds (such as Phorasrhacus longissima *from Patagonia, Argentina, about 17 Ma old) thrived when South America was an island (skull about 50 cm long).*

By about 6 Ma ago South America had drifted close enough to North America for animals to start crossing between the two landmasses. From 3 Ma ago the two landmasses were permanently joined, leading to major exchanges of land animals.

Although animals moved in both directions, most of the South American species were displaced by new species that evolved from North American ones. Marsupial carnivores, flightless carnivorous birds and ground sloths all became extinct. However, a few South American species that migrated northwards, such as the Virginian opossum and nine-banded armadillo, thrive today in North America.

MARSUPIAL RAFT

It is not just marsupials that make Australia's wildlife unique; many Australian plants (fig 113), insects, amphibians, reptiles and birds are also unlike those found anywhere else in the world. Marsupials are thought to have evolved in North America about 110 Ma ago and arrived in Australia after migrating via South America and Antarctica when the landmasses were joined.

About 120 Ma ago Australia drifted away from Antarctica, isolating the Australian species. Elsewhere in the world the placental mammals gradually displaced the marsupials. But in Australia the monotremes, or egg-laying mammals, and the marsupials, thrived and became

adapted to the many different habitats occupied by placental mammals in other continents. Kangaroos and wallabies are grazers that occupy a similar niche to antelope and deer elsewhere, while koalas are sluggish browsers similar to the tree sloths of South America.

The marked differences between the wildlife of Australia and that of Asia were first noted by Alfred Russel Wallace in 1858; the boundary between the two groups is called the Wallace Line (fig 112). He correctly proposed that the two continents had separate evolutionary histories, even though he could not have known at that time that Australia had only relatively recently (25 Ma ago) drifted so close to Asia.

Fig 113 *Pressed specimen and watercolour of* Banksia coccinea, *a distinctive Australian flower.*

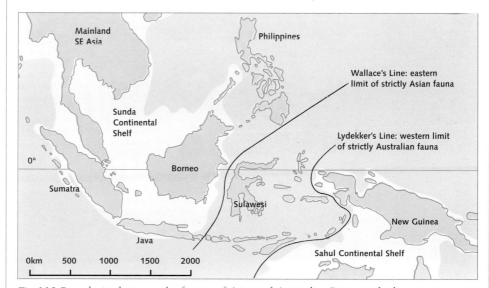

Fig 112 *Boundaries between the faunas of Asia and Australia. Between the lines is a transition zone (Wallacea) with many endemics.*

ON THE MOVE

BRIDGE OR BARRIER?

At the beginning of the Tertiary, 65 Ma ago, the Tethys Ocean stretched from the Atlantic region to the Pacific (fig 137). Over the next 50 Ma, continents to the south (Arabia, India and Australia) collided with Asia, closing the Tethys Ocean, and leaving only the Mediterranean as its last vestige.

India collided with Asia 45 Ma ago, and the ensuing uplift of the Himalaya changed global climate patterns and created the modern monsoon system of Asia (pages 52–53). The northern margin of the Australian plate collided with the south-east margin of the Asian plate 25 Ma ago, creating most of the modern landmasses of the Indonesian archipelago. These provided new and varied habitats, both on land and in the sea, for plants and animals from neighbouring regions to enter and colonize. This, combined with the continually changing and complex pattern of islands, has also given rise to numerous new species. The biodiversity of this region is now one of the richest on Earth, a consequence of its geological history and its present warm, humid climate.

About 25 Ma ago, the collision of Arabia with south-west Asia in what is now the Middle East, completed the closure of the Tethys Ocean. Although intermittent shallow seaways existed for a further 5–10 Ma, most shallow tropical marine plants and animals living in the seas on each side of the new land barrier (Indo-Pacific and Atlantic-Mediterranean) soon became distinct and have remained so ever since. But while separating marine species (fig 114), the formation of this land bridge also enabled land animals and plants to cross between Africa and Asia.

The Earth has been cooling gradually for the last 55 Ma, but the present Ice Age is defined from the time ice had formed at both of the Poles, about 2.5 Ma ago, through to the present day. Although greenhouse gases exert the greatest control over climate, the

Fig 114 Tropical marine life has been evolving separately in the Atlantic and Indo-Pacific, since severance of seaways between them, about 25 Ma ago. Neither of these common reef corals illustrated occurs in the region of the other: table Acropora coral (left) in the Celebes Sea off Malaysia, and elkhorn Acropora coral (right), in the Cayman Islands, Caribbean.

movement of landmasses also has a significant effect (pages 52–53). More recent continental shifts have altered air and ocean currents around the Poles, and triggered the current Ice Age.

When the positioning of landmasses around the globe allows the sea to circulate freely between the Equator and the Poles, warm and cold water mix keeping the oceans at a fairly constant temperature. But if the arrangement of the landmasses restricts ocean circulation, there is less mixing and ice sheets may form around the Poles. When North and South America joined, air and sea current patterns were diverted. Warm ocean currents flowed towards the North Pole, causing increased rainfall over the polar region, which led to the formation of ice sheets.

WARM AND COLD INTERVALS

Although the current Ice Age began 2.5 Ma ago and still continues today, during this time span the climate has fluctuated between glacial and interglacial periods. This is due to cyclical changes in the Earth's orbit (pages 58–59). During a glacial, climate is extremely cold, large ice sheets cover much of the land area outside the tropics (figs 115, 116), and the sea level is low. During an interglacial, climate is warmer, ice is confined to polar regions and high altitudes, and low-lying land is flooded due to rising sea level. We are currently experiencing an interglacial that began about 10,000 years ago so, although there are ice caps over the Poles and ice covers the highest mountains, the rest of the Earth has a much warmer climate (fig 117).

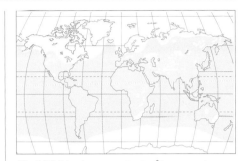

Fig 116 Maximum extent of ice covering the globe about 20,000 years ago, during the last glacial (see fig 118).

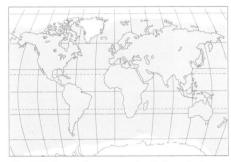

Fig 117 Area of ice over the Poles, during our present day interglacial.

Fig 115 The clearest evidence for former glaciers and ice retreat, are glacial deposits like moraines (left) and till, and gouges on smooth rocks (right) made by glacier-borne boulders (Morteratsch Valley, near St. Moritz, Switzerland).

SHELL THERMOMETER

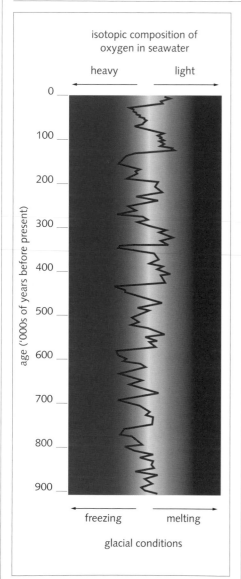

isotopic composition of
oxygen in seawater

heavy → ← light

age ('000s of years before present)

0
100
200
300
400
500
600
700
800
900

← freezing   melting →

glacial conditions

Fig 118 Proportions of light and heavier oxygen in seawater, measured from fossilized shells over the last 900,000 years, reveal alternating periods of warmth and cold.

We know that the Earth is in an ice age, but how do we estimate the continuing fluctuations in temperature over the last million years? Tiny shells of chambered unicellular organisms called foraminifera (fig 119) record the proportions of different atomic types (isotopes) of oxygen in their surrounding seawater. This makes it possible to use fossil foraminifera as thermometers of the past.

When water evaporates, more lighter oxygen atoms are removed from the water than heavier ones. However, the higher the temperature, the higher the proportion of heavier atoms that are also turned into vapour. The cooler the climate, the more heavy oxygen the seas contain, while during times of warmer climates and seas, there is a higher proportion of light oxygen.

Marine animals use the carbon and oxygen in seawater to build their calcium carbonate shells which, when they fossilize, create a permanent record of the proportions of light and heavier oxygen in the sea at that time. When the animals die, their shells accumulate in layers of ocean sediments, enabling scientists to estimate temperature changes in the past.

The oxygen isotope proportions in fossilized shells of different ages have been plotted (fig 118), revealing that over the last 750,000 years, cold intervals (lasting about 90,000 years), have been punctuated by warmer spells of about 10,000 years. In order to understand why these climatic fluctuations have occurred, we need to look at the changes in the Earth's orbit and rotation (see opposite).

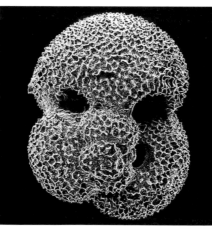

Fig 119 Globigerinoides is one of the commonly used foraminifera for obtaining oxygen isotope measurements for seawater temperatures.

Fig 120 Milankovitch rythms in mid-Cretaceous marls and limestones, Veigon, France.

WANDERING AND WOBBLING EARTH

Small changes in the Earth's rotation and orbit around the Sun affect the amount of heat reaching the Earth's surface. These variations trigger freezing or melting phases within an ice age and could explain why our climate has fluctuated regularly over the last million years (fig 118). The variations occur in three kinds of cycles (fig 121):

a) the shape of the Earth's orbit around the Sun stretches then bulges in cycles of 100,000 and 400,000 years, altering the amount of solar heat the Earth receives;

b) the tilt of the Earth's rotational axis is continually shifting over a cycle of 41,000 years, affecting the distribution of heat over the Earth's surface;

c) the Earth's rotational axis wobbles over a period of about 19,000-23,000 years, changing the distribution of heat over the Earth's surface.

These variations in Earth's orbit are called Milankovitch cycles, after the scientist who first calculated them in 1941. The cyclical patterns coincide with temperature changes estimated from the oxygen isotope pattern (fig 118) and are thought to have caused them. Milankovitch cycles have also been shown to have caused rythmic sedimentation through much of the geological record (fig 120).

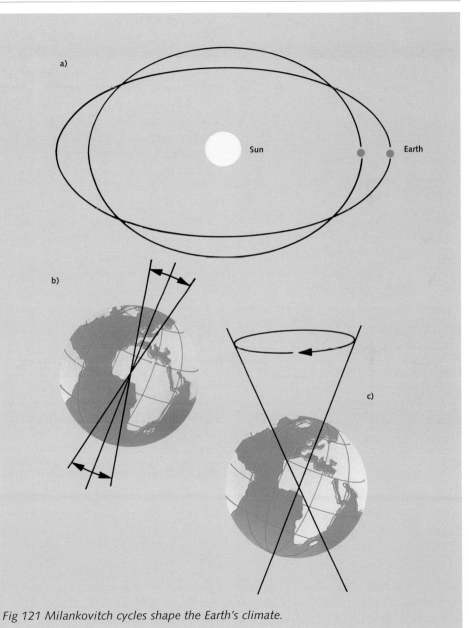

Fig 121 Milankovitch cycles shape the Earth's climate.

WHAT ARE THE EFFECTS OF ICE AGES?

Fig 122 Raised beaches, e.g. at Lismore in northwest Scotland, indicate that relative sea level was once much higher than the current level.

SEAS RISE AND FALL

During an ice age water is locked up in ice, causing sea levels to drop. When climate warms, ice melts, causing sea levels to rise (fig 123). Global changes in sea level, such as those caused during ice ages, are known as eustasy. During recent glacial periods sea level fell as much as 100 m below the present-day level as large volumes of water were bound up in ice sheets. But it is not only the sea level that rises and falls during an ice age; landmasses also sink into the mantle under the weight of large ice sheets and then gradually spring up again when the ice has melted, a process called isostasy (fig 122).

FOSSIL BAROMETER

Fossils are another of the many clues that hint at wide fluctuations in Earth's climate during the last few million years.

Over the last 2.5 Ma many different plant and animal communities have lived in Britain (fig 124). Similar communities tend to return when similar climatic conditions occur. During part of the last interglacial, from 125,000 to 115,000 years ago, a warm maritime climate supported hippos, straight-tusked elephants, narrow-nosed rhinos, fallow deer, badgers, hedgehogs and moles. However, during the early part of the last glacial, from 75,000 to 64,000 years ago, a cold maritime climate supported a very different community of reindeer, bison, wolves, brown bears, red foxes, Arctic hares and northern voles. Communities similar to this are found in Alaska and Arctic Canada today.

Fig 123 Fossilized tree roots exposed at low tide on a beach at Marros Sands, Wales, are the remains of a submerged forest (at least 15,000 years old), that existed when sea level was much lower than it is at present. The wood is preserved by seawater.

Fig 124 This unusual combination of a woolly mammoth jaw 40,000 years old (Mammithus primigenus) with an uyster shell growing on it, was dredged up from the bottom of the North Sea. It reveals that animals once roamed across land that is now flooded by the sea.

As we rapidly approach the present, after hurtling through the thousands of millions of years of Earth's past, we come to the single most important event that has shaped Earth's recent past and is guiding the planet's immediate future – the evolution of ourselves. Although we have already made a massive impact on the Earth and its resources, we have only just emerged in Earth's long and eventful history.

Dryopithecus (fig 125) a probable ancestor of great apes and humans lived a mere 10 Ma ago. It probably looked like an ape with long arms, big hands and no tail, and is likely to have been able to walk semi-upright.

Our earliest-known relatives, around the time of the evolutionary split between apes and humans, are 6-7 Ma old, and their remains have been discovered in Kenya, Ethiopia and Chad. Their successors, the australopithecines, evolved over 4 Ma ago and had an ape-like body, short legs and long arms. Fossilized footprints uniquely preserved in volcanic ash demonstrates that they walked upright.

Our first probable ancestor which combined the human characteristics of a large brain, long legs and the ability to make tools, evolved nearly 2 Ma ago.

Modern humans didn't evolve until perhaps 200,000 years ago from an African ancestor and had replaced Neanderthals (*Homo neanderthalensis*) in Europe by about 30,000 years ago.

The movement of early humans out of Africa followed the intense cooling that triggered the last Ice Age over 2.5 Ma ago. They evolved different body proportions, to survive changing climates as they spread across the globe. Body proportion can be calculated by comparing the length of the shinbone with that of the thighbone. Using these ratios, the body proportions of early humans have been estimated from fossilized bones (see pages 62–63).

Fig 125 Reconstruction od Dryopithecus, *a probable ancestor of great apes and humans.*

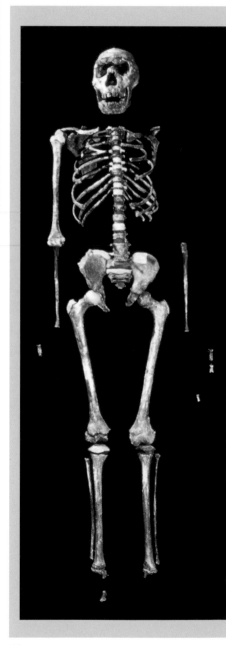

BODY PROPORTIONS

Hot build

Homo ergaster, living in Africa about 1.5 Ma ago, was tall with long slender limbs and narrow hips (figs 126, 129a). This shape maximizes the surface area of the skin, helping to cool the body more quickly in hot dry climates by radiation and sweating.

Cool build

As humans moved north from Africa they would have encountered colder temperatures, greater seasonality and long winters. The first known Britons, *Homo heidelbergensis*, lived on the south coast during an interglacial period about 500,000 years ago. Remains of these early people have been discovered at Boxgrove in Sussex (fig 127, 129b), and include a shinbone. The dimensions of the shinbone indicate that the people were tall but stocky, and able to endure temperate-boreal climates.

Fig 126 The skeleton (1.6 m) of a boy about 10 years old, discovered near Lake Turkana in Kenya, is the most complete Homo ergaster skeleton ever found. It is estimated that he would have grown to about 1.8 m and weighed about 70 kg had he reached adulthood.

Fig 127 Shinbone (30 cm long – specimen not complete) of Homo heidelbergensis from Boxgrove, England.

Cold build

Neanderthals, *Homo neanderthalensis*, lived in Europe and western Asia from over 200,000 years ago until about 30,000 years ago. Measuring the body proportions of Neanderthal remains, scientists have calculated that they were short and wide with stocky limbs – a body shape adapted to cold climates (figs 128, 129c). This shape reduces surface area and helps to conserve heat.

Keeping warm today

Modern humans, *Homo sapiens*, living in Africa and Israel about 100,000 years ago, were tall with long limbs, a body shape adapted to hot climates. It is only from about 20,000 years ago that human remains in Europe have body proportions similar to Europeans today. Humans have now colonized most land surfaces on Earth. Our ability to greatly modify our own environment enables us to survive climatic extremes and other adverse conditions.

Fig 128 Burial site of Neanderthal male, Homo neanderthalensis, *about 60,000 years old, Kebara, Israel (area covered by remains 60x40 cm).*

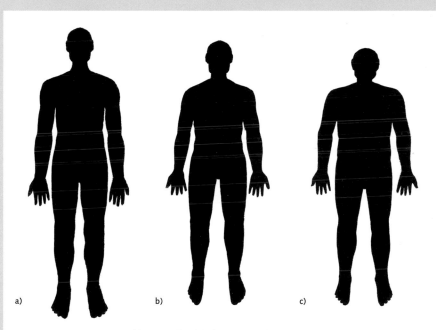

a)   b)   c)

Fig 129 Comparison of human builds based on fossil examples: a) hot, Homo ergaster; *b) cool,* Homo heidelbergensis; *c) cold,* Homo neanderthalensis.

HUMANS SHAPE THE ENVIRONMENT

Fig 130 Moa (Pachyornis elephantophus) *about 5000 years old, South Island, New Zealand.*

As users and makers of tools, humans have used their unique intelligence to take this ability to astoundingly complex levels of technology. Combining our social and technical skills, we have been able to modify and control our own environment, increasingly to the detriment of other species, Earth's resources and even each other. Hunting, farming, wars and industry have all taken their toll.

Our unique use of language to communicate enables us to pass on knowledge through the generations and across different communities. The ability to speak had not evolved in early humans. Although there is no direct evidence, it is assumed that Neanderthals could speak. And there is abundant evidence that from around 30,000 years ago modern humans spoke to each other, traded and used art to express themselves. From the Global Positioning System to videophones and the Internet, humans today are continually developing new technologies to extend our communication networks.

HUNTING

As human populations increased, moved across the globe and became more skilled at hunting, they posed a greater threat to animals hunted for food and skins. The earliest Britons, whose 500,000-year-old remains have been unearthed at Boxgrove in Sussex (fig 127), were skilled at making stone tools that they used to hunt and butcher animals. Many of the remains of larger animals, such as rhinos, horses and deer, bear cut marks made by handaxes – evidence that these early humans were preparing and eating meat.

From about 25,000 years ago humans have been the main cause of the extinction of many big mammals and flightless birds in Australia, North America, Madagascar and New Zealand (fig 131). In North America mammoths and ground sloths became extinct about 10,000 years ago, after humans had spread across the country. Moas, large flightless birds unique to New Zealand, were hunted to extinction (over several hundred years) after humans arrived about 1000 years ago (fig 130).

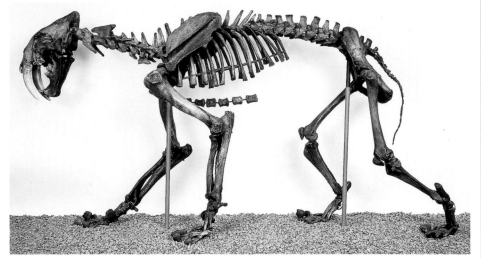

Fig 131 Smilodon fatalis, *a sabre-tooth cat probably hunted to extinction about 14,000 years ago. It was about the size of a present day lion.*

FARMING

Early human societies were hunter-gatherers, constantly on the move to find new food sources. Such societies waned with the gradual advent of agriculture and longer-term settlement, for which there is clear evidence at least 10,000 years ago, perhaps up to 30,000 years ago. Farming was an important change because it provided a regular food source – cereal crops and domesticated animals. But it also transformed the landscape as forests were cleared to harvest timber, raise livestock and plant crops (fig 132). Over half the original area of tropical rainforests has already been cleared for human use.

Rainforest destruction is currently causing the extinction of up to 27,000 species every year. At this rate, at least 10% of all species on Earth could be eliminated over the next 30 years. Are we causing the next mass extinction?

INDUSTRY

During the last 250 years, fossil fuel burning and forest destruction have tripled carbon dioxide levels in the atmosphere, probably increasing the greenhouse effect. More recently, release of industrial chlorofluorocarbons has created a hole in the ozone layer above Antarctica, increasing the amount of harmful ultraviolet radiation entering the atmosphere.

If increasing levels of greenhouse gases cause the predicted rise in temperature there will be devastating effects on the environment – deserts will expand and melting ice will flood low-lying land. However, according to Milankovitch cycle predictions (page 59), the Earth is also due to start cooling as it enters another glacial phase in the next few thousand years. This may be too little and too late however to offset human-induced climate changes.

Fig 132 Hand-cultivated fields in Chimborazo province, Ecuador.

OLD-FASHIONED OATS

A population explosion 7000–8000 years ago coincided with the introduction of pottery – the process of firing clay to produce heat-resistant pots. This enabled Stone Age people to eat cooked liquid cereals such as porridge or gruel. By analysing the microscopic pits and scratches on the surface of teeth from Stone Age people, scientists can discover what sort of food they were eating. Teeth from early Stone Age people have deep abrasions caused by chewing on raw cereal grains. But teeth from late Stone Age people have wear patterns similar to people today, indicating that cooked cereals were already forming a significant part of their diet.

Cooked cereals are more digestible than raw grains because the starch is already partly broken down and they can be eaten more quickly. More calories can be consumed faster, leaving more time available for other tasks. Porridge could also be given to young infants to wean them at an earlier age, possibly leading to larger families. There was a significant population explosion at around this time, but whether cultivation and cooking of cereals caused this, or was an innovative response to it, is still unclear. Moreover, while larger families have more hands for work, denser communities can also stretch resources, increase drudgery and foster disease.

EARTH'S FUTURE

We have panned across Earth's colourful and extensive past in order to understand why our living planet is like it is today. But what of its future, what will it be like, and how and when will it end? Over the next few hundred million years the Earth will re-shape itself again and again as continents continue to glide and collide, mountains rise and fall, and oceans open and close. Scientists are unable to predict these movements much further than about 50 Ma into the future, but during that time the African Rift Valley might widen, and a new ocean may eventually form, with eastern Africa splitting off to the east. The rest of Africa will continue to move north, the Atlantic Ocean will widen and the Pacific will become narrower.

HUMAN HAVOC

The human population has more than doubled in the last 50 years and at current rates is estimated to be rising by 147 people every minute, exerting more and more pressure on the planet's limited resources (fig 133). Although the first humans emerged just 2 Ma ago, a mere blip in geological time, we now dominate most habitats on Earth. Our actions are already polluting and destroying the environment (figs 134, 136), escalating extinction rates, and probably changing climate. The question is whether or not we allow the current scale of destruction to continue and how much it will affect the long-term future of the Earth and its occupants. Will we eventually obliterate all life, including ourselves, or will we learn to sustain our environment and ourselves?

IMPACT FROM SPACE

Earth and the Moon have been repeatedly bombarded by asteroids and comets throughout their history but most impacts are not big enough to cause global effects. A massive impact 65 Ma ago probably contributed to the end-Cretaceous mass extinction. A smaller asteroid or comet less than 100 m in diameter is believed to have entered the atmosphere over the Tunguska region of Siberia in 1908. The resulting shockwave knocked down trees for hundreds of square kilometres. Fortunately, impacts of this scale are extremely rare; an asteroid large enough to cause severe regional or global damage occurs perhaps once every 100,000 years and a comet strike of a similar scale occurs just once every 500,000 years.

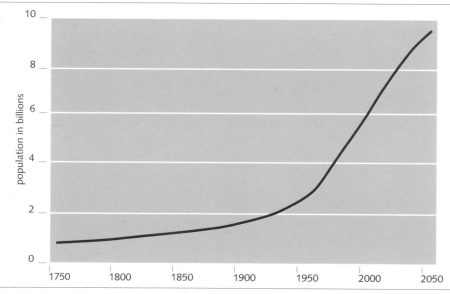

Fig 133 Estimated growth of human population.

Fig 134 Road and fire damage in central Borneo.

Although unlikely, if in future a vast asteroid or comet, bigger than 1–2 km in diameter, ploughed into the Earth, it would trigger a global catastrophe. Dust and gases entering the atmosphere would block out sunlight and cause temperatures to fluctuate wildly and tidal waves may occur. The long-term climatic disruption would destroy food crops and it is possible that all life could be extinguished. However, as space technologies advance, the risk of such an impact is reduced because it is becoming increasingly possible that, if detected in time, a large body heading for Earth could be diverted from its path, thereby preventing worldwide disaster.

DEATH OF STAR

All stars eventually die and the Sun will be no exception. It is estimated that it is currently about half-way through its life cycle, so has enough fuel to last a further 5000 Ma. When all the hydrogen and helium has been exhausted, the star will expand, becoming a red giant. It will reach up to 200 times its present size and will engulf most of the Solar System in flames as far out as the orbit of Mars (fig 135). Eventually it will shrink to roughly the size of the Earth, becoming a dense white dwarf. The smouldering debris will cool and darken and finally be dispersed in space, adding to the cosmic cocktail from which new stars and galaxies will form.

Fig 136 Pollution from an old steel mill, Bihar, India.

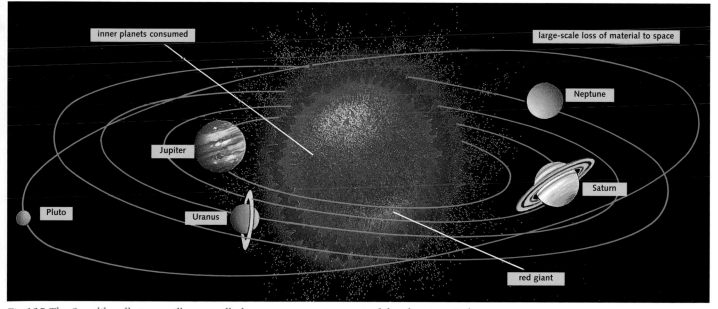

Fig 135 The Sun, like all stars, will eventually burn up, consuming most of the planets as it does.

EARTH THROUGH TIME

■ 'greenhouse' world

■ 'icehouse' world

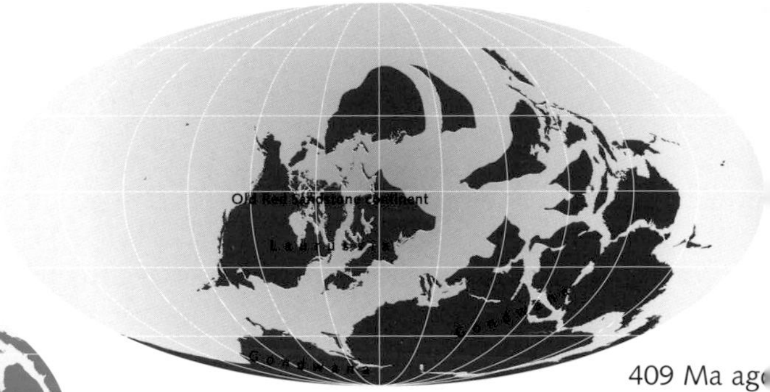

409 Ma ago

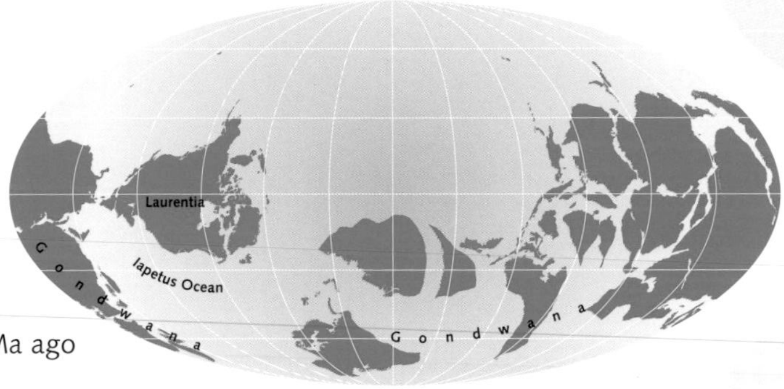

Old Red Sandstone Continent

L a u r u s s i a

G o n d w a n a

525 Ma ago

Laurentia

Iapetus Ocean

G o n d w a n a

G o n d w a n a

340 Ma ago

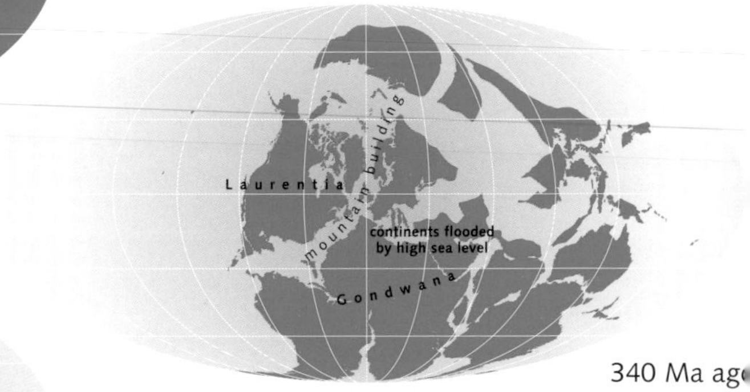

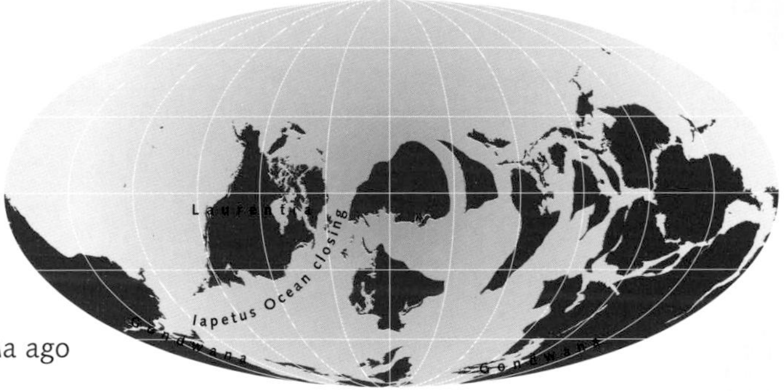

L a u r e n t i a

mountain building

continents flooded
by high sea level

G o n d w a n a

470 Ma ago

L a u r e n t i a

Iapetus Ocean closing

G o n d w a n a

G o n d w a n a

248 Ma ago

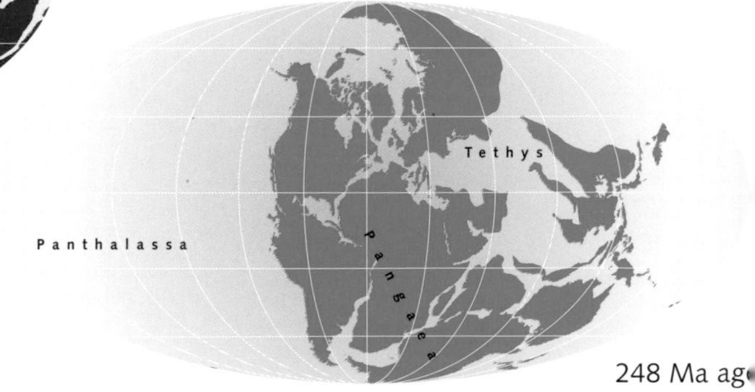

T e t h y s

P a n t h a l a s s a

P a n g a e a

*Fig 137 The globe maps on these pages show how the continents
have glided across the Earth's surface over the last 550 Ma, at
times colliding to form bigger landmasses then later breaking
apart into smaller landmasses. The piecing together of this moving
jigsaw has greatly enhanced our understanding of the rocks and
fossils beneath our feet, and of the present-day distributions of
plants and animals.*

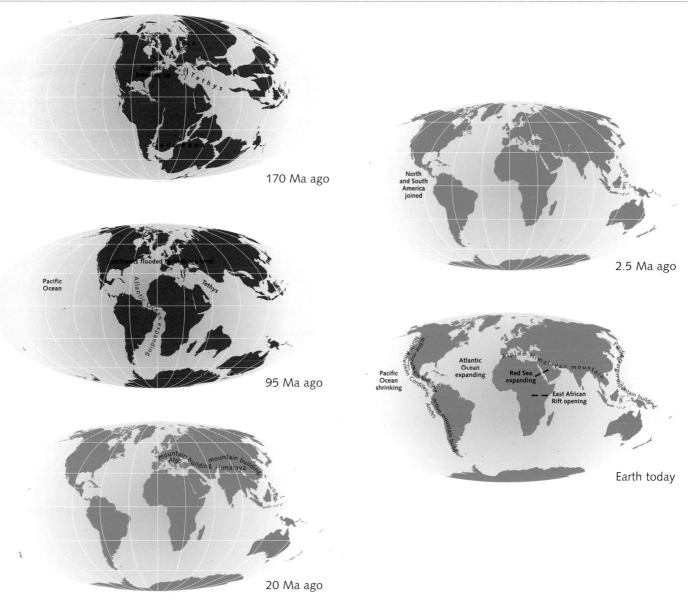

170 Ma ago

Pangaea breaking up

Tethys

95 Ma ago

Pacific Ocean

Continents flooded by high sea level

Atlantic Ocean expanding

Tethys

20 Ma ago

mountain building Alps

mountain building Himalaya

2.5 Ma ago

North and South America joined

Earth today

Pacific Ocean shrinking

active mountain building North American Cordilleras

active mountain building Andes

Atlantic Ocean expanding

Alpine Himalayan mountain

Red Sea expanding

East African Rift opening

Pacific margin mountain building

GLOSSARY

Absolute dating Methods of calculating the exact age of a rock or fossil in numbers of years since its formation, e.g. by measuring the rate of decay of radioactive *isotopes*. cf. *relative dating*.

Aerobic respiration The use of oxygen to convert food into energy, producing carbon dioxide and water as waste products.

Amniotic egg Egg produced by reptiles and birds consisting of a waterproof capsule, protected by an outer shell, containing the embryo bathed in fluid and nourished by yolk.

Asteroids Small planets or fragments of planets in orbit around the Sun in a band between Jupiter and Mars.

Big Bang Model for origin of the Universe about 15,000 Ma ago, when a tiny fireball of infinite density and heat exploded.

Cold-blooded Animal whose body temperature varies according to its level of activity and external temperature. cf. *warm-blooded*.

Continental crust A light, coarse-grained granitic layer, relatively rich in silica, about 35 km thick, forming most of the continental landmasses. cf. *oceanic crust*.

Crustal cycle The process in which *oceanic crust* is constantly being produced (*seafloor spreading*) and reabsorbed (*subduction*) into the *mantle*, which results in the break-up and rearrangement of pieces of *continental crust*.

Eukaryotes Organisms composed of large complex cells with a *nucleus* containing genetic material. cf. *prokaryotes*.

Eustasy Worldwide change in sea level, leading to either increased exposure of seabed or increased drowning of land surfaces.

Evolution Progressive genetic changes in organisms, leading to the emergence of new species.

Geological column The division of time from the formation of the Earth into clearly defined periods, based on *relative* and *absolute dating* of rocks.

Glacial Colder phase within an *ice age*. cf. *interglacial*.

Greenhouse effect The trapping of heat reflected from the Earth's surface within an atmosphere containing carbon dioxide and other greenhouse gases, thus keeping the planet warm.

Greenhouse phase Period in Earth's past when climate was warmer than today with little or no ice around the Poles, due to a more intense *greenhouse effect*. cf. *icehouse phase*.

Half-life The time it takes for half of the radioactive *isotopes* in a mineral to change into stable isotopes.

Hydrothermal vent A gaseous vent where heated fluid rich in water vapour and minerals escapes from the Earth's crust. These are often called black smokers when they occur on the deep ocean floor.

Ice age A cold period in the Earth's history when ice covers much more of the surface, including ice sheets around the Poles.

Icehouse phase Period in Earth's past when climate was cool, with ice sheets around at least one of the Poles, due to a weakening of the *greenhouse effect*. cf. *greenhouse phase*.

Igneous rocks One of the three main rock types, formed when *magma* cools and solidifies. cf. *metamorphic* and *sedimentary rocks*.

Interglacial Warmer phase within an *ice age*. cf. *glacial*.

Isostasy The tendency of the Earth's crust to maintain an equilibrium position in the *mantle*, e.g. slowly rising up after the removal of ice cover.

Isotopes Different forms of the same element that have different numbers of electrons.

Law of superposition Concept, attributed to William Smith, in which as *sedimentary rocks* form, the oldest are at the bottom and the youngest are at the top. cf. *relative dating*.

Magma Hot, molten rock from the Earth's *mantle* and lower crust, which crystallizes to form *igneous rock*, either at the surface (as lava) or below it.

Mantle Interior layer of the Earth, below *oceanic* and *continental crusts* and above the core.

Mass extinction When an abnormally large number of species become extinct worldwide at the same time.

Metamorphic rocks One of the three main rock types, formed when other rocks are altered by heat or pressure. cf. *igneous* and *sedimentary rocks*.

Meteorite Natural object that has survived its fall to Earth from space.

Mid-ocean ridges Large underwater mountainous ridge and rift system on the ocean floor where new *oceanic crust* is produced. cf. *seafloor spreading*.

Milankovitch cycles Three forms of variation in Earth's rotation and its orbit around the Sun that cause changes in climate in cycles of tens to hundreds of thousands of years.

Monsoon Seasonal change in prevailing wind directions

Nucleus Body containing a package of genetic material found in all eukaryotic cells. cf. *eukaryotes* and *prokaryotes*.

Oceanic crust A dense crystalline layer of basaltic composition about 6 km thick, forming most of the ocean floor. cf. *continental crust*.

Ozone layer Form of oxygen found in the Earth's atmosphere.

Photodissociation The splitting of water vapour into hydrogen and oxygen by *ultraviolet radiation* from the Sun.

Photosynthesis The process by which green plants capture energy from the Sun to convert carbon dioxide and water into carbohydrates, releasing oxygen as a waste product.

Planetesimals Mini-planets of varying sizes, up to several thousands of kilometres in diameter.

Prokaryotes Single-celled organisms, such as bacteria, composed of small simple cells whose genetic material is not enclosed in a *nucleus*. cf. *eukaryotes*.

Reducing atmosphere An atmosphere in which oxygen is lost from, or hydrogen added to, compounds.

Relative dating Method of estimating the age of a rock by comparing its position to other rock layers and using distinctive features of the rocks themselves, notably their fossil content, sedimentary features, structure and composition. cf. *law of superposition*.

Seafloor spreading The movement of new *oceanic crust* away from *mid-ocean ridges* where it is formed. cf. *subduction*.

Sedimentary rock One of the three main rock types, formed by breakdown of pre-existing rocks or from organic remains, or precipitation from bodies of surface waters (seas, oceans, rivers, lakes). cf. *igneous* and *metamorphic rocks*.

Silicate rocks Rocks composed mainly of silica, a compound of silicon and oxygen, and one of the most abundant compound.

Subduction The process which occurs along the boundary where two moving tectonic plates are in collision, and where the *oceanic crust* of one plate travels downwards (subduction zone) into the *mantle* beneath the other plate, as part of the *crustal cycle*. cf. *seafloor spreading*.

Ultraviolet radiation Invisible radiation of slightly shorter wavelength and higher energy than the violet end of the visible light spectrum. It is harmful to living tissue.

Unconformity A surface of erosion or non-deposition, usually the former, which separates younger rocks from older rocks.

Uniformitarianism The fundamental concept that geological processes operate consistently through time, so that knowledge of present processes can be used to understand the past.

Warm-blooded Animals that use energy to maintain a constant body temperature regardless of external environment. cf. *cold-blooded*.

Zone fossil A fossil species that characterizes a zone (or sub-zone), the smallest units of fossil-based relative time.

FURTHER READING AND CREDITS

GENERAL READING

Astronomy. S. & J. Mitton. Oxford University Press, new edn., 1996.

The book of life; an illustrated history of the evolution of life on earth. Stephen Jay Gould (general ed.). W. W. Norton, new edn., 2001.

Earth atlas. S. van Rose & R. Bonson. Dorling Kindersley, 1994, out of print.

Earth's restless surface. D. Janson-Smith & G. Cressey. The Natural History Museum, London, 1996.

Evolution. Colin Patterson. The Natural History Museum, London, 2nd edn., 1999.

Fossils; the key to the past. Richard Fortey. The Natural History Museum, London, 3rd edn., 2002.

Gaia; the practical science of planetary medicine. James Lovelock. Gaia Books Ltd., 1991.

The historical atlas of the Earth; a visual exploration of the Earth's physical past. R. Osborne & D. Tarling (eds). Henry Holt & Co. Inc., 1996.

Life; an unauthorised biography; a natural history of the first four thousand million years of life on Earth. Richard Fortey. Harper Collins, London, 1997.

Meteorites. Sara Russell & Monica Grady. The Natural History Museum, London, 2nd edn., 2002.

Search for Life. Monica Grady. The Natural History Museum, London, 2001.

Volcanoes. S. van Rose & I. Mercer. The Natural History Museum, London, revd. 2nd edn., 1999.

TEXT BOOKS

Earth and life through time. S.M. Stanley. W.H. Freeman & Company, New York, 1986, out of print.

Historical geology; evolution of the Earth and life through time. R. Wicander & J.S. Monroe. Brooks/Cole, 4th edn., 2004.

Holmes' principles of physical geology. P.McL.D. Duff. Chapman & Hall, 4th edn., 1993.

Mass extinctions and their aftermath. A. Hallam & P.B. Wignall. Oxford University Press, 1997.

An outline of Phanerozoic biogeography. A. Hallam. Oxford University Press, 1994.

Phanerozoic sea-level changes. A. Hallam. Columbia University Press, New York, 1992.

The young earth; an introduction to Archaean geology. E.G. Nisbet. Allen & Unwin, London, 1987.

PICTURE CREDITS

Unless listed below all photographs are copyright the Natural History Museum, London (NHM)
Key: BC = Bruce Coleman Ltd, PP = Panos Pictures, SPL = Science Photo Library
Fig 1 Corbis-Bellmann/UPI
Front cover, inside front cover and figs 3, 5, 6, 7, 8, 15 NASA
Fig 19 Mary Gee and Euan Nisbet, Royal Holloway University of London
Fig 24 and 43 Caroline Jones
Fig 25 SPL/B. Murton/Southampton Oceanography Centre
Fig 26 SPL/Tony Craddock
Fig 27 SPL/NASA
Fig 29 Pamela Reid, University of Miami
Fig 30 Timothy Miller, University of Maine
Fig 32 Geoscience Features
Fig 33 Dr. B. Booth/Geoscience Features
Fig 34 A.M. Page, Royal Holloway, University of London
Fig 39 Andrew Milner
Fig 40 BC/George Bingham
Figs 42, 57, 67, 99, 100, 115 Brian Rosen, NHM
Fig 85 PP/Rob Huibers
Fig 86 Dr. Mike Widdowson, Research Fellow, Open University
Fig 88 PP/Chris Sattleberger
Fig 89 BC/Christer Fredriksson
Fig 98 F. Xavier Valldeperas, Universitat Autonoma de Barcelona
Fig 104 V.L. Sharpton, Lunar and Planetary Institute, Houston, Texas
Fig 108 André Matula
Fig 110 BC/Dr. P Evans
Fig 114 Linda Pitkin, NHM
Fig 120 Andy Gale, NHM
Fig 122 Courtesy of The British Geological Survey © NERC
Fig 123 Kevin Church, Open University
Fig 126 National Museum of Kenya
Fig 128 The Kebara Archive
Fig 132 PP/Jeremy Horner
Fig 134 PP/Tatlow
Fig 136 PP/Paul Smith
Illustrations by: Ray Burrows (Figs 60, 61, 76, 90, 103, 106, 129), Mike Eaton (Figs 2, 4, 28, 35, 47, 49, 66, 78, 112, 116, 117, 121, 135), Gary Hincks (Figs 21, 44, 65, 87, 107), Perks Willis Design (Figs 13, 14, 22, 38, 41, 83, 84, 118, 133, 137)
Fig 78 Adapted from p.108, *Prehistoric life.* David Norman. Boxtree, London, 1994.

ACKNOWLEDGEMENTS

This book would not have been possible without help from many colleagues at the Natural History Museum, London too numerous to list individually. While the book is intended for non-specialist readers, it draws on recently published research including work by many Museum colleagues.

Brian Rosen would like to thank his family for all their support and interest throughout, and his colleagues for their patience.

Front cover design: The Design Studio, NHM
Design: David Robinson
Editor: Karin Fancett
Picture research: Emily Hedges
Printed by: Craft Print, Singapore

Distributed in Australia and New Zealand by
CSIRO Publishing
PO Box 1139
Collingwood, Victoria
Australia
ISBN 0-643-09069-X

Facing page: Reconstruction of scene at a tar pool at Rancho la Brea. The mammoth has become trapped in the oily waters, and is being devoured by the sabretooth cat, Smilodon.

Back cover: Fossil bony fish, Pristigenys substriatus Eastman, preserved in fine-grained limestone from Mont Bolca, northern Italy, about 40 Ma old (length of specimen, 17 cm).

THE HENLEY COLLE